# Electronic Image Display

Equipment Selection and Operation

# Electronic Image Display
## Equipment Selection and Operation

**Jon C. Leachtenauer**

SPIE PRESS
A Publication of SPIE—The International Society for Optical Engineering
Bellingham, Washington USA

Library of Congress Cataloging-in-Publication Data

Leachtenauer, Jon C.
 Electronic image display : equipment selection and operation / Jon C. Leachtenauer.
   p.  cm. – (SPIE Press monograph ; PM 113)
 Includes bibliographical references and index.
 ISBN 0-8194-4420-0
 1. Information display systems. I. Title. II. Series.

TK7882.I6L43 2003
621.3815'422—dc21                                                                    2003054304


Published by

SPIE—The International Society for Optical Engineering
P.O. Box 10
Bellingham, Washington  98227-0010 USA
Phone: (1) 360.676.3290
Fax: (1) 360.647.1445
Email: spie@spie.org
Web: www.spie.org


Copyright © 2004 The Society of Photo-Optical Instrumentation Engineers

All rights reserved. No part of this publication may be reproduced or distributed
in any form or by any means without written permission of the publisher.

Printed in the United States of America.

*To my wife, Mary Ellen, who has patiently awaited completion of this project, and to Amy, Caroline, Paul, Jon, and Eleanor.*

# Contents

| | | |
|---|---|---|
| **Preface** | | xiii |
| **Acknowledgments** | | xv |
| **List of Acronyms** | | xvii |

**Chapter 1  Introduction** ............................................................. 1
    1.1  The Image Chain ............................................................. 1
    1.2  The Display as a System ............................................... 5
    1.3  Characterizing Quality ................................................... 6
    1.4  Reader Road Map ........................................................... 7
    References ............................................................................ 8

**Chapter 2  Measurement of Light and Color** ........................... 11
    2.1  Light Measures ............................................................... 11
    2.2  Light Measurement ........................................................ 15
    2.3  Color Measures ............................................................... 17
    2.4  Color Measurement ........................................................ 26
    2.5  Summary .......................................................................... 26
    References ............................................................................ 27

**Chapter 3  Electronic Display Operation** .................................. 29
    3.1  Display Types .................................................................. 29
    3.2  Display Controller .......................................................... 31
    3.3  CRT Operation—Monochrome .................................... 34
    3.4  CRT Operation—Color ................................................. 40
    3.5  The AMLCD .................................................................... 41
    3.6  Plasma Displays .............................................................. 43
    3.7  Display Controls ............................................................. 44
        3.7.1  Luminance controls .......................................... 44
        3.7.2  Geometry controls ............................................ 46
        3.7.3  Color controls ................................................... 48
    3.8  Summary .......................................................................... 49
    References ............................................................................ 50

**Chapter 4  Physical Display Quality Measures** ........................ 53
    4.1  Resolution Measures ...................................................... 54
        4.1.1  Addressability and screen size ....................... 54
        4.1.2  Pixel density and size ...................................... 55

|  | | 4.1.3 | Pixel subtense | 57 |
|---|---|---|---|---|
|  | | 4.1.4 | Resolution-addressability ratio | 59 |
|  | | 4.1.5 | Edge sharpness | 59 |
|  | | 4.1.6 | Contrast modulation | 60 |
|  | | 4.1.7 | Raster modulation | 62 |
|  | | 4.1.8 | Modulation transfer function | 63 |
|  | | 4.1.9 | Bandwidth | 64 |
|  | 4.2 | Contrast Measures | | 64 |
|  | | 4.2.1 | Bit depth | 64 |
|  | | 4.2.2 | Dynamic range | 66 |
|  | | 4.2.3 | Gamma | 66 |
|  | | 4.2.4 | Input/output function | 67 |
|  | | 4.2.5 | Halation | 67 |
|  | | 4.2.6 | Reflectance and transmittance | 68 |
|  | | 4.2.7 | Luminance stability | 69 |
|  | | 4.2.8 | Luminance and color uniformity | 69 |
|  | | 4.2.9 | Gamut | 70 |
|  | | 4.2.10 | Viewing angle | 71 |
|  | 4.3 | Noise Measures | | 71 |
|  | | 4.3.1 | Signal-to-noise ratio | 71 |
|  | | 4.3.2 | Noise power spectrum, noise-equivalent quanta, and detective quantum efficiency | 73 |
|  | | 4.3.3 | Jitter, swim, and drift | 74 |
|  | | 4.3.4 | Refresh rate and flicker | 74 |
|  | | 4.3.5 | Warm-up and aging | 74 |
|  | 4.4 | Artifacts and Distortions | | 74 |
|  | 4.5 | Categorization by Measurement Domain | | 77 |
|  | 4.6 | Summary | | 79 |
|  | References | | | 79 |
| **Chapter 5** | **Perceptual Quality and Utility Measures** | | | **81** |
|  | 5.1 | Subjective Quality Ratings | | 83 |
|  | 5.2 | Subjective Performance (Utility) Estimates | | 84 |
|  | 5.3 | National Imagery Interpretability Rating Scale | | 85 |
|  | 5.4 | Objective Perceptual Quality Measures | | 87 |
|  | | 5.4.1 | Briggs target | 89 |
|  | | 5.4.2 | Briggs vs. NIIRS | 95 |
|  | 5.5 | Objective Performance (Utility) Measurement | | 96 |
|  | | 5.5.1 | Theory of signal detection | 99 |
|  | | 5.5.2 | Time measures | 102 |
|  | 5.6 | Summary | | 102 |
|  | References | | | 102 |

| | | |
|---|---|---|
| **Chapter 6** | **Performance of the Human Visual System** | **105** |
| | 6.1 Physiology of the HVS | 105 |
| | 6.2 Visual Performance | 111 |
| |     6.2.1 Separable acuity | 111 |
| |     6.2.2 Stereo acuity | 117 |
| |     6.2.3 Color vision performance | 117 |
| | 6.3 Individual Differences | 119 |
| | 6.4 Models of Visual Performance | 124 |
| |     6.4.1 Monochrome (luminance) models | 124 |
| |     6.4.2 Color models | 133 |
| | 6.5 Summary | 134 |
| | References | 134 |
| **Chapter 7** | **Contrast Performance Requirements** | **137** |
| | 7.1 Performance Requirements | 137 |
| | 7.2 Measurement Definition | 138 |
| | 7.3 Requirement Rationale | 139 |
| | 7.4 Instrument Measurement | 148 |
| |     7.4.1 Initial setup | 149 |
| |     7.4.2 Dynamic range | 149 |
| |     7.4.3 Lmax | 150 |
| |     7.4.4 Input/output function | 150 |
| |     7.4.5 Luminance uniformity | 150 |
| |     7.4.6 Viewing angle threshold | 152 |
| |     7.4.7 Halation | 153 |
| |     7.4.8 Bit depth | 153 |
| |     7.4.9 Color temperature | 154 |
| |     7.4.10 Color uniformity | 155 |
| | 7.5 Measurement Alternatives | 155 |
| | 7.6 Summary | 158 |
| | References | 158 |
| **Chapter 8** | **Size and Resolution Performance Requirements** | **161** |
| | 8.1 Performance Requirements | 161 |
| | 8.2 Measurement Definition | 162 |
| | 8.3 Requirement Rationale | 162 |
| | 8.4 Instrument Measurement | 170 |
| |     8.4.1 Screen size (diagonal) | 170 |
| |     8.4.2 Screen aspect ratio | 170 |
| |     8.4.3 Pixel aspect ratio | 170 |
| |     8.4.4 Addressability | 171 |
| |     8.4.5 Pixel density | 171 |
| |     8.4.6 Contrast modulation—Zone A | 172 |
| |     8.4.7 Contrast modulation—Zone B | 172 |

|  |  |  |
|---|---|---|
| | 8.5 Measurement Alternatives | 173 |
| | 8.6 Summary | 178 |
| | References | 178 |

**Chapter 9  Noise, Artifact, and Distortion Performance Requirements  181**

    9.1 Performance Requirements    181
    9.2 Measurement Definition    181
    9.3 Requirement Rationale    183
    9.4 Instrument Measurement    188
        9.4.1 Warm-up time    188
        9.4.2 Scan rate    189
        9.4.3 Jitter, swim, and drift    189
        9.4.4 Macro and micro jitter    190
        9.4.5 Luminance step response    190
        9.4.6 Moiré    191
        9.4.7 Extinction ratio    192
        9.4.8 Mura and other artifacts    192
        9.4.9 Pixel defects    193
        9.4.10 Signal-to-noise ratio    193
        9.4.11 Straightness (waviness)    193
        9.4.12 Linearity    194
    9.5 Measurement Alternatives    194
    9.6 Summary    197
    References    197

**Chapter 10  Monitor Selection and Setup  199**

    10.1 Monitor and Video Controller Selection    199
    10.2 Monitor Setup    203
        10.2.1 Monitor connection and setup    203
        10.2.2 Controlling the monitor environment    204
        10.2.3 Monitor calibration    209
        10.2.4 Perceptual linearization    210
    10.3 Display Maintenance    214
    10.4 Summary    216
    References    216

**Chapter 11  Pixel Processing  219**

    11.1 Pixel Intensity Transforms    219
        11.1.1 Dynamic range adjustment    219
        11.1.2 Tonal transfer adjustment/correction    220
        11.1.3 Color transforms    223
    11.2 Spatial Filtering    224
    11.3 Geometric Transforms    227
    11.4 Bandwidth Compression and Expansion    229

|  |  |  |
|---|---|---|
| | 11.5 Sequence of Operations | 233 |
| | 11.6 Summary | 234 |
| | References | 234 |

**Chapter 12  Digitizers, Printers, and Projectors** — **237**

    12.1 Digitizers — 237
        12.1.1 Digitizer operation — 237
        12.1.2 Digitizer image quality and device selection — 239
        12.1.3 Digitizing procedures — 243
    12.2 Printers — 248
        12.2.1 Printer operation — 249
        12.2.2 Printer quality and selection — 249
        12.2.3 Printing procedures — 254
    12.3 Projection Displays — 258
    12.4 Summary — 259
    References — 259

**Appendix:  Test Targets** — **261**

**Index** — **265**

# Preface

This book provides guidance on maintaining image quality in the selection and operation of electronic displays. The book is intended for anyone who must perform critical information extraction tasks using electronically displayed continuous-tone imagery, particularly in medical and military applications. It is also of value to managers and operations and maintenance personnel associated with such tasks, as well as supporting procurement personnel. The book is written at multiple levels such that a variety of users can find the information needed to perform their jobs. At a minimum, the individual user can determine how to select and evaluate a viewing system. For those readers interested in proceeding further, the rationale for recommendations is provided, using both image examples and results of empirical studies. Five of the chapters cover the fundamentals of display operation, the human visual system, and image quality measurement. Measurement procedures are provided for those readers who have access to measurement instrumentation, and alternatives are provided for those without such access. A CD is included that contains a wide range of test targets.

The book begins with an overview and examples demonstrating the importance of maintaining image quality in the display process. The display chain is defined and briefly reviewed. A road map for readers with differing needs is provided. Chapter 2 introduces light and color measures and measurement. Chapter 3 provides a brief overview of electronic display operation. Both CRT and flat-panel display technologies are covered, although the emphasis is on CRT technology. The operation of common display controls is demonstrated with graphs and image examples.

Chapters 4 and 5 discuss physical and perceptual display quality measures. Physical measures include measures of resolution, contrast, and noise, both spatial and temporal. Perceptual measures are rating scales and performance measures used to rate the absolute or relative perceived quality of a display. Chapter 6 provides information on the performance of the human visual system. A brief description of the physiology of the eye is followed by a discussion of visual system capabilities—spatial, contrast, and color. The effects of individual differences are also described (including aging effects). The chapter ends with a review of visual performance models, with emphasis on the Barten model used as the basis for the NEMA/DICOM display calibration process. Subsequent chapters draw on the literature using these measures to illustrate the effects of display quality parameters.

The next three chapters of the book (7–9) provide guidance in display selection, covering luminance and spectral measures, resolution measures, and temporal/spatial measures. Each section begins with a listing of the recommended performance parameters and criteria values for both monochrome and color displays. The parameters are defined, the selection criteria are provided, and the measurement procedures are described at both the perceptual and physical levels. Sources of performance information and their interpretation are discussed. Results of studies on key quality measures are provided where available. These studies are drawn from both the surveillance/reconnaissance and medical literature. Numerous figures are provided showing both measurement definitions and image examples to illustrate the effect of the key quality measures. Many of the desired performance measures are not routinely provided by vendors and require sophisticated equipment for measurement. Equipment and measurement procedures are defined for organizations that have either the capability of acquiring and operating such equipment or of specifying measurement performance requirements to vendors. For individuals or organizations without such capabilities, simplified procedures and tools are provided. Many of the tools are perceptual.

The operating environment is a critical factor in maintaining image quality. Recommended procedures are provided in Chapter 10 with emphasis on the control of room lighting. Chapter 10 also covers monitor selection, setup, and maintenance. Monitor luminance compensation techniques to account for the performance of the human visual system and procedures for generating the necessary look-up table are described. Monitor performance degrades with age, so the effects of the aging process are explained. Procedures for periodic quality assessment are defined.

Since software manipulation of an image is an important part of the image chain, Chapter 11 covers pixel processing operations including tonal, color, spatial filtering, and geometric manipulation. The proper sequence of operations is defined and alternative methods of processing discussed. A final chapter provides guidance on hard-copy capture and presentation. Digitizer properties are described, and guidance on digitizer selection and operation is provided. The process of transferring displayed soft-copy images to presentation media such as prints and transparencies is discussed. Printer calibration and look-up table generation procedures are defined to best emulate the originally displayed image on the presentation media. A brief section on electronic projection displays is also included.

**Jon C. Leachtenauer**
*September 2003*

# Acknowledgments

I have learned from many people over my career. My work in the field of display technology began at the Boeing Company. I would particularly like to thank Dr. Jim Briggs, Mr. John Booth, Mr. Richard Farrell, Dr. Conrad Kraft, Dr. Charles Elworth, and Mr. Richard Schindler.

The idea for this book came from two projects I worked on at the National Exploitation Laboratory (NEL) and the National Imagery and Mapping Agency (NIMA). Both projects were directed at developing guidelines for the display of imagery on soft-copy displays. I was fortunate to work with the staff of the National Information Display Laboratory (NIDL) and Eastman Kodak Company. I would like to thank Mr. Michael Grote, Dr. Ron Enstrom, Mr. Michael Brill, Mr. Albert Pica, Dr. Jeff Lubin, and Dr. Dennis Bechis of the NIDL. I would also like to thank Mr. John Mason, Mr. Matt Pellichia, and Mr. Jim Leuning of Kodak. At NIMA, I would like to thank Mr. Art Cobb.

I deeply appreciate the valuable editing help supplied by Ms. Ellen Schwartz. Review comments provided by Dr. Peter Barten of the Barten Consultancy and Dr. James Florence of ELCAN Optical Technologies were invaluable. I would also like to acknowledge Ms. Margaret Thayer of SPIE, who has been very helpful in guiding me through the publication process.

Finally, I am grateful for the help and support of many other people that have contributed to my knowledge and understanding over the past 45 years. In particular, the staff of NIMA and predecessor organizations have made my career a rewarding experience.

# List of Acronyms

| | |
|---|---|
| ACM | alternating-current matrix |
| ACR | American College of Radiology |
| AFC | alternative forced choice |
| AMLCD | active-matrix liquid crystal display |
| ASICS | application specific integrated circuits |
| CCD | charge coupled device |
| CD | compact disc |
| CD ROM | compact disc, read-only memory |
| CIE | Commission Internationale d'Eclairage (International Commission on Illumination) |
| CL | command level |
| Cm | contrast modulation |
| CMYK | cyan/magenta/yellow/black |
| CPU | central processing unit |
| CR | computed radiology |
| CRT | cathode ray tube |
| CSF | contrast sensitivity function |
| CT | computed tomography |
| CTF | contrast transfer function |
| DAC | digital-to-analog converter |
| DCS | dynamic color separation |
| DCT | discrete cosine transform |
| DICOM | Digital Imaging and Communication in Medicine |
| DPCM | delta pulse code modulation |
| dpi | dots per inch |
| DQE | detective quantum efficiency |
| DR | dynamic range |
| DRA | dynamic range adjustment |
| DROC | differential receiver operating characteristic |
| DSIS | double-stimulus impairment scale |
| DSCQS | double-stimulus continuous quality scale |
| EIA | Electronic Industries Association |
| FED | field emissive display |
| FFT | fast Fourier transform |
| FOV | field of view |
| FROC | free response operating characteristic |
| GSD | ground-sampled distance |
| HDTV | high-definition television |
| HSB | hue/saturation/brightness |

| | |
|---|---|
| HVS | human visual system |
| I/O | input/output |
| IDEX | Image Display and Exploitation |
| IEC | integrated exploitation facility |
| IT | information technology |
| JND | just-noticeable difference |
| JPEG | Joint Photographic Experts Group |
| LCD | liquid crystal display |
| Lmax | maximum luminance |
| Lmin | minimum luminance |
| LUT | look-up table |
| MPEG | Motion Pictures Experts Group |
| MRI | magnetic resonance imaging |
| MTF | modulation transfer function |
| MTFC | modulation transfer function compensation |
| NC | noise criterion |
| NEMA | National Electrical Manufacturers Association |
| NEQ | noise-equivalent quanta |
| NIDL | National Information Display Laboratory |
| NIIRS | National Imagery Interpretability Ratings Scale |
| NIMA | National Imagery and Mapping Agency |
| NIST | National Institute of Standards and Technology |
| NPS | noise power spectrum |
| NS | not (statistically) significant |
| NTSC | National Television Systems Committee |
| OLED | organic light-emitting diodes |
| PAC | picture archiving and communications |
| PACS | picture archiving and communications system(s) |
| PC | personal computer |
| PDP | plasma display panel |
| PM | photomultiplier |
| ppi | pixels per inch |
| RAM | random access memory |
| RAR | resolution addressability ratio |
| RER | relative edge response |
| RGB | red/green/blue |
| ROC | receiver operating characteristic |
| SAR | synthetic aperture radar |
| SCS | sequential color separation |
| SMPTE | Society of Motion Picture and Television Engineers |
| SNR | signal-to-noise ratio |
| SQS | subjective quality scale |
| STN | supertwisted nematic |
| TN | twisted nematic |
| TSD | theory of signal detection |

| | |
|---|---|
| TTA | tonal transfer adjustment |
| TTC | tonal transfer correction |
| UCS | uniform chromaticity spacing |
| USAF | United States Air Force |
| UV | ultraviolet |
| VESA | Video Electronics Standards Association |

# Chapter 1
# Introduction

The last 30 years have seen a steady migration from film-based imagery to electronic display of imagery. This migration has occurred in the medical community, the reconnaissance and surveillance community, and in the publication industry. It has been accompanied by a proliferation of advances in display technology. Unfortunately, these advances have not always improved the conveyance of image information. Further, the wide variety of display technology now available and rapid changes in the marketplace make the selection of a display increasingly difficult. Are flat-panel displays superior to the classic CRT? Will plasma displays make LCDs obsolete? What is the difference between a flat CRT and a flat-panel display? Will FEDs replace LCDs? We explore these issues and many others in this book.

## 1.1 The Image Chain

The electronic display can be considered as an element in an image chain (Fig. 1.1) whose purpose is to capture and convey information in a form most useful to the human observer. Each element of the chain, including the observer, affects the conveyance of information. The chain begins with a capture device, which acquires data using energy in some portion of the electromagnetic spectrum ranging from x rays to radio waves. The capture device may form an electronic image as with a digital camera, or it may simply acquire a file of intensity and position as with, for example, computed tomography. Regardless of the original capture device, the information must be displayed in a form useful to a human observer. Intensity information outside the visible spectrum must first be processed to convey variations in light or color. For example, radiographs code variations in x-ray tissue density as shades of gray; multispectral scanners code variations in reflectivity as variations in color. Beyond the basic coding of intensity relative to position, there are certain image manipulations that can be performed to enhance the conveyance of information. They range from simple transforms in intensity space to complex neural network processing algorithms designed to at least partially replace the human observer in the detection process. The scope of this book is limited to film digitizers as a capture device and to basic pixel processing transforms.

The main thrust of this book is on the electronic display element of the image chain. Although the electronic display is the primary medium used to extract information from an image, hard-copy imagery is still frequently used to convey the extracted information to others because prints and viewgraphs can be annotated

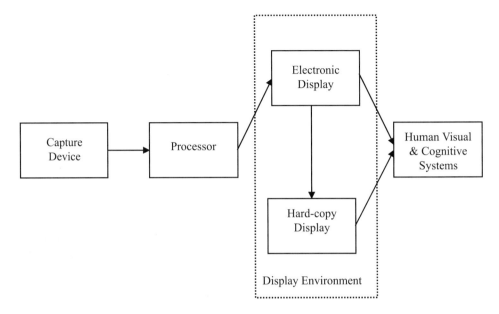

**Figure 1.1** Image chain.

and displayed to large groups. Thus, this book discusses digital image printers and printing operations. But electronic projection devices are becoming more common (and affordable) as replacements for hard copy; therefore, this book also covers these devices as an element in the image chain.

A display is designed (hopefully) to convey image information. We can measure the ability of a display to convey information in both the physical and the perceptual realms. For example, a display's contrast ratio and resolution are measured with the implicit assumption that more is better. It will be shown that this is not always the case. We can also apply various perceptual and task measures to evaluate the performance of a display. The reconnaissance and surveillance community relies on task-based scales such as the National Imagery Interpretability Rating Scale (NIIRS);[1] the medical community performs detection studies based on the theory of signal detection (TSD).[2] Ideally, a set of physical measures would predict perceptual performance, but this is not yet the case—we can understand the physical parameters that affect performance but cannot today model their relative contributions.

The performance of an electronic display is strongly affected by the environment in which it is operated, particularly the surrounding light level. With film display systems (light tables and light boxes), room lighting is relatively unimportant because the strong light sources used to illuminate the film generally overpower any effects of room lighting. However, electronic displays are much more limited in their intensity, and the output of the display can easily be affected by the room lighting. With text, the quality of a display is rather obvious—ultimately, the observer can either read the text or cannot. With imagery, performance quality may be less obvious because the observer may not know that details are missing. Figures 1.2 and 1.3 show examples to illustrate this point. The image on the left in Fig. 1.2 was taken under bright room lighting conditions, the image on the right under darkened conditions. The numbers are more legible on the image at the right. Figure 1.3 illustrates portions of an MRI scan with the same relationship as the images in Fig. 1.2. Differences are not apparent, and it is possible that detail in the "dark" image might be lost in the "bright" image.

The final element in the chain is the human observer. The human visual and cognitive (brain) systems act together to determine whether or not the desired information is conveyed. The visual system has a limited capability to detect small

**Figure 1.2** Effect of room lighting on text image appearance.

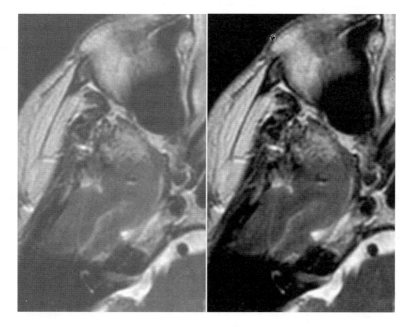

**Figure 1.3** Effect of room lighting on appearance of MRI scans.

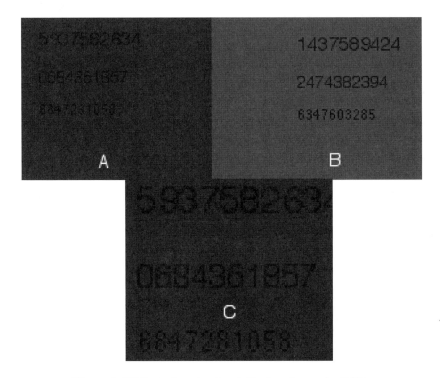

**Figure 1.4** Effect of contrast and display size on legibility.

spatial detail and low-contrast information. Figure 1.4 illustrates this point. The numbers on the image in A are of a size and contrast that makes them nearly illegible. The contrast has been increased in B and the numbers are legible. In C, the contrast is the same as in A, but the size of the numbers has been increased to make them legible. On typical displays, low-contrast detail is too small to be seen by the observer, so either the image must be enlarged or the contrast must be increased. Text is typically viewed at very high contrast (black on white) so small details can be seen. With imagery, the contrast is often much lower so magnification is required.

The ability to detect light intensity differences is not linear with display intensity variations. Unless the display is matched to the visual system, information is lost. One technique, called "perceptual linearization," applies a look-up table to better match the display output to the observer.[3] But the brain does not always respond to physical differences as we might expect, particularly with color. Further, the ability to detect features in an image is a function of both the visual process and the thought process. Finally, not all observers are alike, and our visual performance degrades with age. We need to understand this degradation and learn how to compensate for the loss.

## 1.2 The Display as a System

Although we may think of a display as the device on which we view an image, we can also consider it as a system (Fig. 1.5). The display system is comprised of a storage and processing device, a display controller or video card, and the display itself. The quality and usefulness of the displayed image are affected by all three components. The processor (and its hosted software) transforms the image into a form that best matches the requirements of both the display and the observer. As discussed previously, the eye cannot see all of the detail present in an image on a typical display. The processor thus enlarges or magnifies the image using a transform that minimizes information loss. The processor also applies other transforms that are designed to maximize information conveyance using contrast adjustment, noise reduction, and edge sharpening. Thus the processor can, in some cases, make up for limitations of the display. The sequence in which the processing is applied has a significant impact on the final appearance of the image. The processor software can also degrade the appearance of images relative to their optimum.

The display controller is the interface between the processor and the display. With conventional CRT displays, the controller transforms digital data to an analog signal that drives the display. The controller also performs a variety of other functions including signal timing, graphics acceleration to speed up the display of imagery, and the application of look-up tables designed to optimize the relationship between digital image values and output light intensity. A poorly performing controller can add noise and various artifacts to images and slow the process of writing new images to the display. Even the physical connection between the processor/controller and display can impact quality. Excessively long connectors and improperly matched signals can produce a variety of image artifacts or anomalies.

The display itself is where we see the visible evidence of image or display quality. The display outputs light or color in proportion to input voltage from the controller. The display also has hardware to vary perceived brightness, contrast,

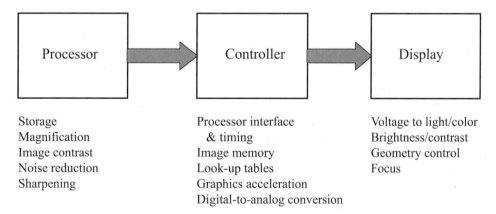

**Figure 1.5** Display system.

and geometric positioning. In terms of the display, quality is a function of the type of display, the display design, and a long list of performance parameters.

## 1.3 Characterizing Quality

The VESA's Flat Panel Display Measurements Standard lists 44 different measures of display quality.[4] The EIA's "Display Measurement Methods under Discussion" lists 29.[5] These measures characterize the ability to see small detail and distinguish small brightness or color differences. However, no single measure or even a simple combination of measures fully characterizes quality. The number of potential measures defies our ability to make rational purchase choices. Compounding the problem is the fact that most vendors supply information on only four or five quality parameters. The popular literature has stopped providing in-depth performance reviews; thus there are few sources of in-depth information on display quality. Display users are largely left on their own to choose from among several vendors and competing technologies. This book addresses the issue.

Each user has a set of tasks to perform using the display system. These tasks can be defined in terms of detection objectives, such as the radiologist seeking to detect pulmonary nodules on a chest radiograph or the military analyst looking for tanks. In both cases, successively more detailed detection is often required. For example, the military interpreter, having detected a tank, seeks to detect features of the tank that will enable tank model identification. At each level of detection, the objects or features have a range of sizes and gray level or color contrast. If the features are presented to the observer with sufficient size, contrast, and sharpness, the detection task can be performed. Figure 1.6 illustrates this concept. The listed display quality measures represent only a small subset of the nearly 50 measures discussed in this book.

The original capture device places fundamental limits on size, contrast, and sharpness—the display system cannot add information. It can process the information to better match the characteristics of the human visual system (HVS) or it can degrade the information that is presented to the display observer. It is this degradation that we seek to avoid in the selection and operation of a display system.

The fundamental quality issue is "Will the display system provide the information I need for a price I can afford?" Although a quality/cost trade-off exists, the optimum point in that trade-off depends on the user's requirements. Thus, this book guides the user through a series of choices in order to select the type of display that will best satisfy the user's requirements. Once the display type is selected, further choices will determine performance—detail size, sharpness, and contrast. Information is provided on the meaning of the many quality measures used in the display literature as well as a more practical subset of those measures. A set of recommended minimum performance requirements is provided, along with instrumented measurement methods. Wherever possible, noninstrument methods are provided for users lacking access to measurement instrumentation.

# INTRODUCTION

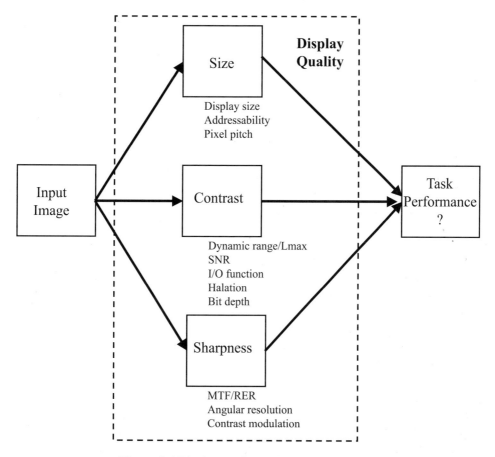

**Figure 1.6** Display quality and task performance.

## 1.4 Reader Road Map

This book is written for a wide variety of users of soft-copy image display systems. It should be emphasized that the focus is on tasks using continuous-tone imagery, both monochrome and color. Recommendations thus do not necessarily apply to office tasks such as word processing or related tasks where the display is typically only two tones (e.g., black and white).

The contents of this book are of special interest to those in the medical and reconnaissance and surveillance fields, but users from other fields with critical image viewing applications will also find useful information. The intended audience ranges from individual display system users to management and facilities personnel involved in the procurement and setup of soft-copy display systems, including information technology (IT) and image science personnel involved in system design and maintenance. Because of the diversity of the intended audience, some readers may prefer to read only the sections most pertinent to their interests.

For the reader new to the field of soft-copy display, Chapter 2 discusses light and color measurement, and Chapter 3 describes display system operation. For readers interested in display quality metrics, Chapter 4 describes physical quality measures and Chapter 5 describes perceptual and cognitive measures. The information in these chapters will enable the reader to readily understand the large body of available display performance literature.

Because the display user is a significant element in the display image chain, Chapter 6 discusses the performance of the HVS. Display users should be aware of the noncorrectable visual performance effects associated with aging. Chapter 6 also addresses models that describe and predict the performance of the visual system.

For those involved in display performance assessment, Chapters 7 through 9 provide measurement procedures for the subset of measures considered most important in defining display performance. Procedures are defined for those users having the necessary instrumentation; alternatives are provided for those users without instrumentation. Detailed information on display quality requirements is also provided in these chapters along with the rationale for each requirement. Chapter 4 presents a more comprehensive list of measures and provides references to additional sources of information on measurement procedures.

The reader interested solely in display system purchasing advice can find it in Chapter 10. Those looking for guidance on selection and operation of scanners and printers are referred to Chapter 12.

Since the quality of imagery displayed on a soft-copy system is dependent on how it is set up and operated, system users, managers, and IT personnel will find Chapters 10 and 11 of special interest. Those engaged in facility design for soft-copy systems will find relevant information discussed in Sec. 10.2.

The CD provided with this book contains a wide variety of test targets that can be used to perform the quality measures described in Chapters 7 through 9. They also provide the basis for comparing the relative quality of alternative display systems and components. Finally, they can be used to measure display performance on a continuing basis as a means of quality control.

## References

[1] J. C. Leachtenauer, "National Imagery Interpretability Rating Scales: overview and product description," *ASPRS/ASCM Annual Convention and Exhibition Technical Papers,* Vol. 1, Remote Sensing and Photogrammetry, American Society for Photogrammetry and Remote Sensing and American Congress on Surveying and Mapping, Baltimore, MD, April 22–25, 1996, pp. 262–271.

[2] C. E. Metz, "Fundamental ROC analysis," in *Handbook of Medical Imaging,* J. Beutel, H. L. Kundel, and R. L. Van Metter (eds.), Vol. 1, Physics and Psychophysics, SPIE Press, Bellingham, WA, pp. 751–770 (2001).

[3] S. M. Pizer, "Intensity mapping to linearize display devices," *Computer Graphics Image Processing*, Vol. 17, pp. 262–268 (1981).

[4] Video Electronic Standards Association, *Flat Panel Display Measurements Standard,* Version 2.0, VESA, Milpitas, CA (2001).
[5] National Information Display Laboratory, *Display Monitor Measurement Methods under discussion by EIA Committee JT-20, Part 1, Monochrome CRT Monitor Performance*, Version 2.0, Princeton, NJ (1995).

# Chapter 2
# Measurement of Light and Color

Although the observer's perception at the end of the display chain is how we assess the quality of a display, it is useful to have a set of quantitative physical measures to describe the performance of the display itself. This chapter describes the fundamental measurement units and the instruments used to measure light and color. Chapters 4–10 will describe the application of these measures.

The terms "light" and "color" are used because that is how we characteristically think of displays and other sources of light energy. As this chapter will show, light energy is measured in physical units but often described in units of perception or vision. Light invokes the perception of brightness; the wavelength distribution of light energy invokes the sensation of color. An understanding of both the physical and perceptual aspects is necessary to understand the literature that describes the performance of displays.

## 2.1 Light Measures

Light refers to the energy in the portion of the wavelength spectrum that can be seen by the HVS (Fig. 2.1).[1] Light is radiant energy in the region between 400 and 700 nm. The distribution of energy as a function of wavelength produces the perception of color.

Objects above a temperature of 0 K emit radiant energy. As temperature increases, the amount of emitted energy increases; above 700–800 K, we can begin to see radiant energy in the visible spectrum, i.e., light. Objects first emit a red color, and as temperature increases, the color moves from red to orange to white.

Radiant energy is measured in units of power. The watt (W) is the primary unit of radiant power. Thus, a 100-W light bulb produces 100 W of radiant power. Not all of this power is radiated within the visible spectrum, however. The measurement of radiant energy is called radiometry. Radiant energy can be measured as a function of wavelength or as measurements integrated over a wavelength band.

The term "photometry" refers to measurements of the total visual spectrum where the measurements are weighted according to the wavelength sensitivity of the visual system. The sensitivity function is shown in Fig. 2.2. Photometry, not radiometry, is of primary interest in this book. Whereas radiometry measures radiant power, photometry measures luminous power. Luminous power is radiant power weighted by the spectral sensitivity of the eye. The primary unit of luminous power is the lumen (lm).

Light arises from a source of energy such as a candle, a light bulb, or the sun. If we are at a sufficient distance from the energy source, we can consider it a point

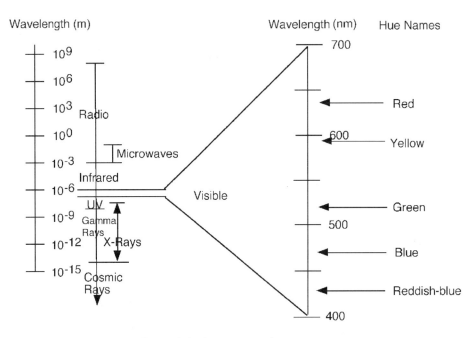

**Figure 2.1** Electromagnetic spectrum.

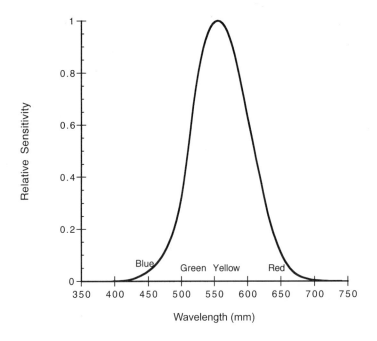

**Figure 2.2** Visual standard sensitivity function for photopic vision. Data from Ref. [2] based on the CIE 1931 observer.

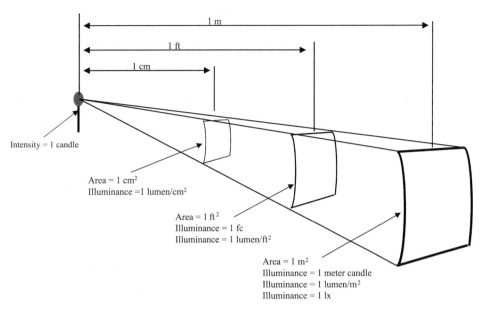

**Figure 2.3** Units of illuminance.

source emitting equally in all directions. If we measure the energy falling on a surface, we are measuring what is called illuminance (Fig. 2.3)[3]. Thus, the lamp illuminates the page of this book, and the energy falling on this page from a lamp or light from a window is measured in terms of illuminance. Figure 2.4 shows the relationship between lighting conditions and illuminance. Illuminance is measured using two units: the lux (lx) and the footcandle (fc). One lux is the luminous energy falling on a one-meter square 1 meter from the candle emitting 1 lm per steradian (sr). A steradian is the measurement of a solid angle enclosing an area on a sphere (surrounding the point source) equal to the square of the sphere's radius. The total area of the sphere is $4\pi$ sr. A footcandle is the luminous energy falling on a one-foot square 1 foot from a candle with an intensity of 1 lm/sr. The conversions between lux and footcandle are

$$lx = 10.764 \text{ fc, and fc} = 0.0929 \text{ lx}. \tag{2.1}$$

In the context of displays, illuminance is used to measure the light coming from sources other than the display. If this light falls on the display, it reduces the contrast of the displayed image. This issue is discussed in more detail in Chapter 10.

With displays, we are more interested in the light coming from (emitted by) the display, i.e., luminance. Luminance is measured in units of candelas per square meter ($cd/m^2$) or foot-lamberts (fL). The conversion is

$$cd/m^2 = 0.292 \text{ fL, and fL} = 3.426 \text{ cd/m}^2. \tag{2.2}$$

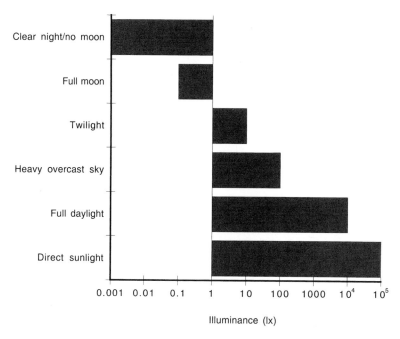

**Figure 2.4** Illuminance as a function of light source. Data from Ref. [9] used with permission of Peter A. Keller.

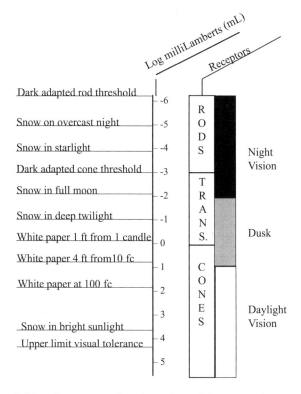

**Figure 2.5** Luminance as a function of condition. Data from Ref. [3].

A unit called the "nit," another name for the candela, is also used, but it is no longer considered proper terminology. The unit milli-lambert (mL) is used occasionally. Conversions are

$$mL = 0.929, \text{ and } fL = 3.183 \text{ cd/m}^2. \qquad (2.3)$$

Luminance is also used to measure the light reflected from objects. In the context of displays, this involves room lighting and glare sources reflected from the screen. If you are wearing a light-colored shirt and viewing a CRT in office lighting, it is likely that you can see the image of the shirt reflected from the screen. Luminance is used to measure this reflected light. Figure 2.5 shows luminance values for a variety of common conditions.[3]

The perception of luminance is called brightness—a display with a high level of luminance is said to be bright. Luminance and illuminance measurements cannot be interchanged except in terms of the passage of light into the eye. This is discussed in Chapter 6.

## 2.2 Light Measurement

Luminance and illuminance relative to displays are normally measured using a photometer. A photometer uses a light-sensitive detector that is usually filtered to represent the response of the eye. An illuminance photometer or meter measures

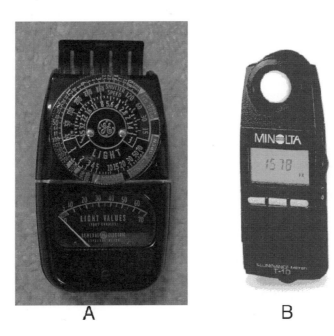

**Figure 2.6** Illuminance photometers. Image A by author; image B courtesy of Minolta Corporation.

the light from a source or the light falling on a surface. A camera exposure meter is a type of illuminance meter. Two types of illuminance meters are available. The typical low-cost photographic light meter (Fig. 2.6A) is pointed at a source and measures the light (generally reflected light) from the source. The field of view (FOV) or angle of acceptance defines the area measured. More expensive illuminance photometers measure and integrate all of the light over a hemisphere located at the point of measurement (Fig. 2.6B). The meter corrects for the effects resulting from differences in the angles of incidence of the light.

Several types of luminance photometers are available. At the low end of the cost spectrum are photometers with measurement pucks that attach by suction to the face of the display or are held in place by the observer against the face of the display. The sensor is shielded such that only light emitted by the display is measured (Fig. 2.7A). Light from the areas surrounding the display (ambient light) that falls on the display is generally eliminated, which may be desirable if only the display output is of interest but may be undesirable if the user must measure both emitted display light and reflected room light together. The latter procedure is recommended for certain display adjustment procedures that require knowledge of the total light coming from the display. Because the puck is held with a suction cup, it is often necessary to tape the puck to the monitor to prevent the puck from slipping, but the tape should not be allowed to touch the face of the monitor. Costs are on the order of $1000 to $2000 (U.S.).

By limiting the acceptance angle of the photometer, either with an aperture or optically, the photometer can be moved away from the display (Fig. 2.7B). Optical

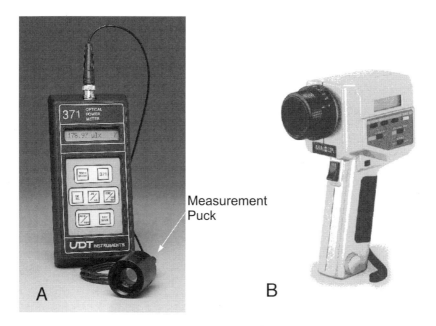

**Figure 2.7** Luminance photometers. Image A courtesy of Belfort Instrument Co. and image B courtesy of Minolta Corporation.

control generally provides a smaller FOV (down to 0.3 deg) but is more costly ($3000 and up). So-called microphotometers are used to generate spot profiles, either by scanning (moving the image or the photometer) or by using a linear detector array called a charge-coupled device (CCD). Finally, CCD arrays can be used to image portions of a display and measure several points at once. With associated software, they can make many of the measurements described in Chapter 4 in a short period of time. Such devices, however, are very costly (> $50,000).

A sample of current manufacturers of luminance measurement devices includes Gamma Scientific, Graseby, Minolta, Monaco, Photo Research, and UDT. Listings of sources can be found in Refs. [4] and [5].

Aside from the issue of contact vs. remote measurements, device selection is also based on the range of luminance values that can be measured and the accuracy of the measurements. For display applications, a range of 0.03 to 900 $cd/m^2$ is desirable. The National Image Display Laboratory (NIDL) recommends 0 to 500 $cd/m^2$ (see Ref. [6]). Absolute accuracy should be better than ±10% over the measurement range. Photometers degrade over time and must be recalibrated to a standard source on at least a yearly basis. Not all devices can be recalibrated, which should be a consideration for the initial purchase. The calibration source should be certified by the National Institute of Standards and Technology (NIST) rather than a secondary source. Several photometer vendors offer calibration services.

## 2.3 Color Measures

Unlike light, where the relationship between energy level and perception has a single dimension, color is considerably more complex. A variety of systems have been developed to describe the perception of color. For example, hue, saturation, and brightness are all used to describe the subjective impression of color, as depicted in Fig. 2.8. Hue is the term used to describe what we normally think of as color; red and green are hues. Saturation describes what we might think of as the intensity of the color, and brightness describes the lightness or darkness of the color; pink is a less saturated red. The Munsell system assigns color names or codes by matching unknown colors against paint chips that are at specific points on the hue/saturation/brightness cylinder.[1] More commonly in the context of displays, color is measured in terms of a color coordinate system. This method is always related to or is some version of the CIE (Commission Internationale de l'Eclairage) coordinate system.

A color coordinate system is based on the fact that any color can be produced from combinations of three primary colors. For displays, where colored light is combined, the primary colors are red, green, and blue. By combining the three primaries in various amounts, all other colors can be produced (Fig. 2.9). This is known as additive color.

Color coordinates are defined by matching "unknown" colors to colors defined by measured amounts of the basic primary colors. The primary colors are defined in terms of wavelength (Fig. 2.10). Working under defined conditions, observers

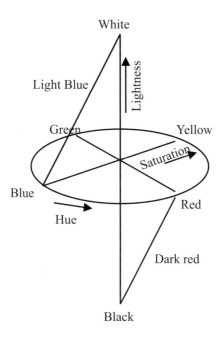

**Figure 2.8** Subjective descriptions of color.

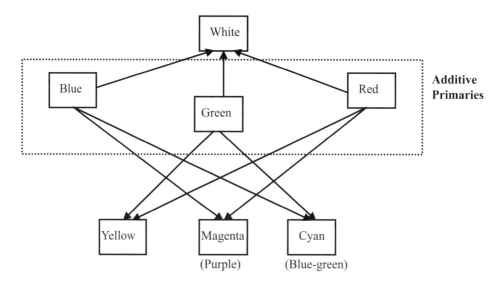

**Figure 2.9** Additive primaries.

match colors to those containing measured amounts of the three primary colors. The measured amounts are called tristimulus values. Since the tristimulus values must sum to 1, color coordinates can be described in terms of two coordinates as shown in Fig. 2.11, which depicts the extent of physically realizable colors as well as the coordinates of specific wavelengths.

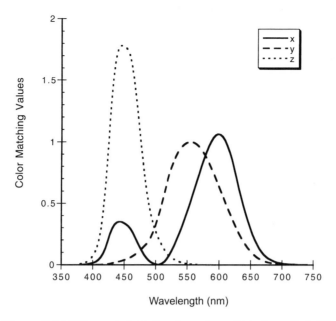

**Figure 2.10** CIE spectral tristimulus values. Data from Ref. [2].

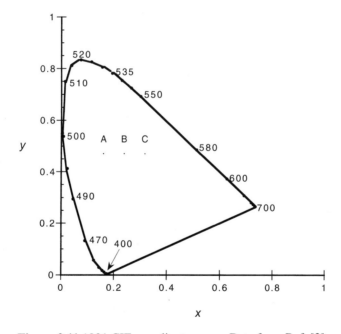

**Figure 2.11** 1931 CIE coordinate space. Data from Ref. [2].

Ideally, one would like the differences between colors to be equal because then any x-y coordinate difference would equate to a discriminability difference. For instance, in reference to Fig. 2.11, one would like the spacing difference between A and B to equal the spacing difference between B and C. This is not the case with the original

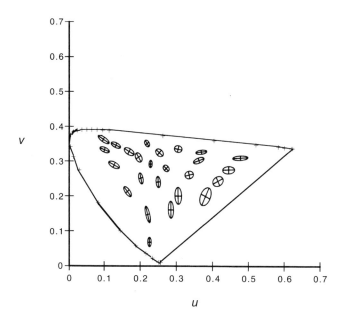

**Figure 2.12** CIE 1960 UCS diagram with MacAdam ellipses. Data from Ref. [2].

CIE diagram, but various transforms have been proposed to achieve this equality. In 1960, the CIE published the Uniform Chromaticity Spacing (UCS) diagram. This diagram (Fig. 2.12) is a transform of the original $x, y, z$ diagram.[2] Although the UCS color space was more uniform than the $x, y, z$ space, it was still far from uniform. The ellipses shown on the figure represent measures of observer variability in color matching. Each ellipse represents 10 times the standard deviation of a single observer attempting to match two colors under ideal conditions. The ellipses, known as MacAdam ellipses, represent the variability in repeated matchings. If the color space were perceptually uniform, all of the ellipses would be circles with the same diameter. The variation in size suggests the nonuniformity of the color space.

A later version of the UCS was published in 1976 with coordinates defined in terms of $u'$ and $v'$.[2] This space is shown in Fig. 2.13.

The CIE color spaces do not account for luminance differences. Three-coordinate systems that are made up of one luminance value plus two color-coordinate values have been designed to account for this lack. Most directly related to displays is CIE's $L^*u^*v^*$ space.[2] This space was developed from a transform of the 1976 $u'v'$ space. The initial transform to $u'v'$ is obtained from the $x, y$ coordinates by

$$u' = \frac{4X}{(X + 15Y + 3Z)} \text{ and}$$
$$v' = \frac{9Y}{(X + 15Y + 3Z)},$$
(2.4)

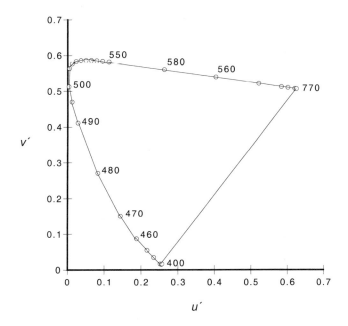

**Figure 2.13** CIE 1976 UCS diagram. Data from Ref. [2].

where the values $X$, $Y$, and $Z$ are transforms of the $x$, $y$, and $z$ chromaticity coordinates. Called tristimulus values, they are designed to avoid negative values and are defined as

$$X = \frac{x}{y}V,$$
$$Y = V, \text{ and} \quad (2.5)$$
$$Z = \frac{z}{y}V.$$

The value $V$ is luminance weighted in accordance with the luminous efficiency function of the visual system.

The $L^*u^*v^*$ space is further defined by

$$L^* = 116\left(\frac{Y}{Y_O}\right)^{1/3} \text{ with } \frac{Y}{Y_O} > 0.008856,$$
$$L^* = 903.29\left(\frac{Y}{Y_O}\right) \text{ with } \frac{Y}{Y_O} \leq 0.008856, \quad (2.6)$$
$$u^* = 13L^*(u' - u'_O), \text{ and}$$
$$v^* = 13L^*(v' - v'_O),$$

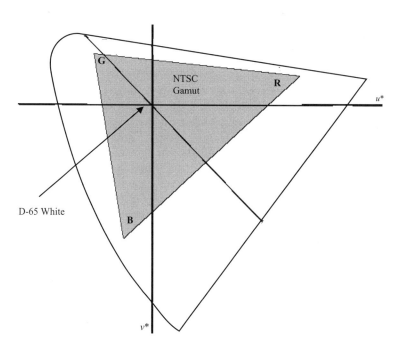

**Figure 2.14** Display color space.

where $Y$ is the measured luminance, $Y_o$ is the luminance of the display white point (the maximum luminance or Lmax value), $u'$ and $v'$ are the transformed coordinates of the measured point, and $u_o'$ and $v_o'$ are the transformed coordinates of the display white point. The display white point occurs when the display is commanded to Lmax. At Lmax, the display appears white. In a color display, the white point occurs when all three colors are at their maximum value. The $u*$ axis passes through (roughly) the red and green point and the $v*$ axis passes through the blue and yellow point. The precise location depends on the coordinates of the white point (Fig. 2.14).[7] The gray area shown in Fig. 2.14 represents the gamut of the display. The gamut is the range of colors that can be realized by a color display (or any device that displays colors such as a color printer). For a color display, the gamut is defined by the coordinates of each color (red, green, and blue, or R, G, B) when that color is commanded to the maximum command level (CL) and the other two colors to 0. The maximum CL is the highest digital value available for the particular display.

The 1976 $L*a*b*$ space is also a transform of the 1931 CIE $x$, $y$ values.[2] The value $L*$ is defined as in Eq. (2.6). The values $a*$ and $b*$ are defined as

$$a* = 500\left[\left(\frac{X}{X_O}\right)^{\frac{1}{3}} - \left(\frac{Y}{Y_O}\right)^{\frac{1}{3}}\right] \text{ and}$$
$$b* = 200\left[\left(\frac{Y}{Y_O}\right)^{\frac{1}{3}} - \left(\frac{Z}{Z_O}\right)^{\frac{1}{3}}\right], \tag{2.7}$$

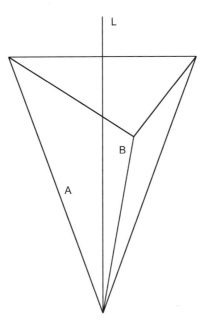

**Figure 2.15** Color space measurement.

where $X_o$, $Y_o$, and $Z_o$ are the $X$, $Y$, and $Z$ coordinates of the illuminant and represent the measured values. The value of $L^*a^*b^*$ space is that it is device independent. It is commonly used in color processing applications such as Adobe Photoshop™.

In both the $L^*a^*b^*$ and $L^*u^*v^*$ spaces, a color difference is defined as the three-dimensional distance between two points (Fig. 2.15) such that

$$\Delta E_{ab} = (\Delta L^{*2} + \Delta a^{*2} + \Delta b^{*2})^{0.5} \quad \text{and} \qquad (2.8)$$
$$\Delta E_{uv} = (\Delta L^{*2} + \Delta u^{*2} + \Delta v^{*2})^{0.5} \;.$$

Color differences can be defined in terms of the perception of "colorfulness." Colorfulness has two dimensions: chroma and saturation.[2] Chroma is defined as

$$C^*_{ab} = (a^{*2} + b^{*2})^{0.5} \quad \text{and} \qquad (2.9)$$
$$C^*_{uv} = (u^{*2} + v^{*2})^{0.5} \;;$$

and saturation is defined as

$$S^*_{ab} = \frac{C^*_{ab}}{L^*} \quad \text{and}$$
$$S^*_{uv} = \frac{C^*_{uv}}{L^*} \;. \qquad (2.10)$$

Equation (2.9) indicates that the perception of colorfulness increases as we move toward the extremes of the color space (red-green, blue-yellow). Equation (2.10) indicates that for a given chroma level, increasing the brightness level decreases the perception of colorfulness.

Color differences can also be described in terms of hue differences. The quantities

$$\Delta H^*_{ab} = \left[ (\Delta E^*_{ab})^2 - (\Delta L^*)^2 - (\Delta C_{ab})^2 \right]^{0.5} \text{ and}$$

$$\Delta H^*_{uv} = \left[ (\Delta E^*_{uv})^2 - (\Delta L^*)^2 - (\Delta C_{uv})^2 \right]^{0.5} \tag{2.11}$$

describe hue differences.

Because they are encountered in software applications, other less rigorous models will be described. One of these models, the RGB model, defines color in terms of the digital CL values of each of the three additive primary colors (R, G, B).[1] The brightest red obtainable in 8-bit space has the RGB value of 255,0,0. As the red becomes darker, the R value decreases and the G,B values remain 0. The brightest possible yellow has an RGB value of 255,255,0. Figure 2.16 illustrates RGB space. In terms of CIE coordinates, the space varies depending on the coordinates of the corner points. There are several versions of RGB space models.[8]

The hue/saturation/brightness (HSB) space shown previously in Fig. 2.8 is also used in some software applications. Values are expressed as angular values from 1 to 360 (representing hue), horizontal distance from the center (indicating saturation), and vertical location on the cylinder (indicating brightness).

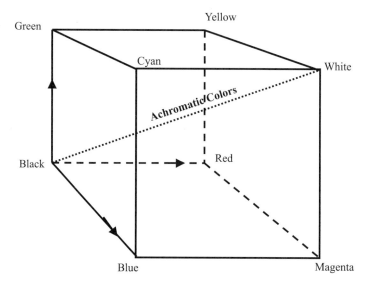

**Figure 2.16** RGB space.[1]

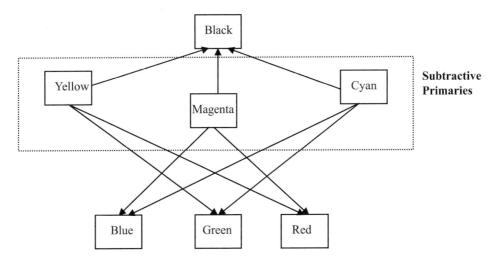

**Figure 2.17** Subtractive color process.

Another color space model is CMYK,[8] which is designed for printing applications. The printing process uses subtractive colors, i.e., the inks absorb colors. The primary colors, CMY, are cyan (a blue-green), magenta (a purple or red-blue), and yellow. Note that these colors are located midway on the axes between the additive primaries of red, green, and blue. With inks, a fourth color (black), denoted by K, is needed to get a true black. Figure 2.17 is a simplified illustration of the subtractive color concept. The subtractive primaries absorb their complementary (opposite) colors; thus, cyan absorbs red, and magenta absorbs green. When cyan and magenta are used together, the only color not absorbed is blue, so the light reflected appears blue. In practice, this is not quite true since the actual colors that can be obtained depend on the nature of the available inks.

To aid in understanding the relationship between some of the color spaces described, Table 2.1 lists HSB, RGB, CMYK, and $L*a*b*$ values for a variety of colors. The colors are shown in the file called colors.tif on the enclosed CD.

**Table 2.1** Color coordinate comparison.

| Color | $L*a*b*$ | HSB | RGB | CMYK |
|---|---|---|---|---|
| Deep red | 52/79/74 | 0°/1/1 | 255/0/0 | 0/84/99/0 |
| Burgundy | 41/64/–5 | 330°/84/73 | 187/29/107 | 22/95/19/6 |
| Orange | 64/48/77 | 27°/98/99 | 253/116/4 | 0/63/99/0 |
| Bright green | 88/–75/87 | 120°/100/100 | 0/255/0 | 55/0/94/0 |
| Bright blue | 32/66/–109 | 240°/100/100 | 0/0/255 | 100/79/0/0 |
| Pink | 80/26/11 | 0°/180/180 | 255/180/180 | 0/38/17/0 |
| Purple | 28/59/–77 | 268°/100/75 | 90/0/190 | 94/89/0/0 |
| Black | 0/0/0 | 0°/0/0 | 0/0/0 | 63/52/51/100 |

## 2.4 Color Measurement

The measurement of colors requires a filtered photometer (also called a colorimeter) or a spectroradiometer. For most applications, only luminance is of interest, the exceptions being color printing applications and display white-point verification.

A filtered colorimeter uses multiple filtered detectors to measure the tristimulus values of $X$, $Y$, and $Z$.[9] The $Y$ measurement equates to luminance and the colorimeter can be used to measure both luminance and color coordinates. The measured values are used to compute the color coordinates in $x$, $y$ or $u'v'$ space. Measurements may also be made in $L*a*b*$, RGB, HSB, or any color space that can be derived from CIE values. Luminance measurement should be accurate to within ±5%.

Colorimeters are offered both as measurement pucks and as aperture-limited remote devices. Costs range from under $1000 (U.S.) for puck devices to nearly $20,000 for remote devices. Note, however, that some low-cost devices may offer only control of the gamma and white point of a display as opposed to providing color coordinate measurement output.

Scanning spectroradiometers scan an area of the display at each wavelength over the visible spectrum.[9] These values can then be converted to any one of the common color-coordinate spaces. Costs range from $10,000 to close to $100,000.

Some current manufacturers of color measurement devices include Gamma Scientific, Hoffman, Minolta, Photo Research, Quantum Data, and X-Rite. Refer to Refs. [4] and [5] for additional sources.

## 2.5 Summary

The light energy emitted from a display is measured in units of luminance. Luminance is measured in candelas per meter squared ($cd/m^2$), foot-lamberts (fL), or milli-lamberts (mL). The term "nit" may sometimes be encountered, for which values are also measured as $cd/m^2$. Light falling on a display is measured in units of illuminance as lux (lx) or footcandles (fc). Luminance and illuminance photometers are used to measure luminance and illuminance. The perception of luminance is called brightness.

Color is measured and described in several two- and three-dimensional spaces. The spaces correspond to the perception of hue, saturation, and brightness. For displays, the CIE $L*u*v*$ space is commonly used. It provides a more perceptually uniform space than other systems but is still not completely uniform. Not all of the colors defined by the CIE space can be realized with a display; the color limits of a display are defined as its gamut. Colorimeters and spectroradiometers are used to measure color. These devices can also be used to measure luminance.

## References

[1] R. J. Farrell and J. M. Booth, *Design Handbook for Imagery Interpretation Equipment*, Boeing Aerospace Co., Seattle, WA (1984).
[2] G. Wyszecki and W. S. Stiles, *Color Science*, Second Edition, John Wiley & Sons, Inc., New York (2000).
[3] J. W. Wulfeck, A. Weiz, and M. W. Raben, *Vision in Military Aviation*, WADC TR 58-399, Wright Air Development Center, Wright-Patterson Air Force Base, OH (1958).
[4] Society for Information Display, "Annual directory of the display industry," *Information Display*, Vol. 16(8), pp. 37–83 (2002).
[5] Photonics Spectra, *The Photonics Buyers' Guide*, Book 2, 48th International Edition, Laurin Publishing, Pittsfield, MA (2002).
[6] National Information Display Laboratory, *Display Monitor Measurement Methods under discussion by EIA Committee JT-20, Part 1, Monochrome CRT Monitor Performance*, Version 2.0, Princeton, NJ (1995).
[7] National Information Display Laboratory, *Evaluation of the Barco MPRD 9651 Monitor*, Princeton, NJ (1994).
[8] H. R. Kang, *Color Technology for Electronic Imaging Devices*, SPIE Press, Bellingham, WA (1996).
[9] P. A. Keller, *Electronic Display Measurement*, John Wiley & Sons, Inc., New York (1997).

# Chapter 3
# Electronic Display Operation

This chapter begins with a description and categorization of those electronic displays typically used to view continuous-tone imagery. The operation of these devices is described in terms of how a captured digital image is transformed into a displayed image on the face of a display. An image is acquired either by a digital imaging system (a digital camera of some type) or by scanning an image that was acquired by an analog device (a conventional camera or x-ray machine). This chapter emphasizes those factors that affect the quality of the displayed image (relative to the original), although quality is not explicitly discussed in this chapter. Quality is defined here as the degree to which a displayed image accurately portrays the original scene. This chapter also describes the operation of the hardware controls that affect the appearance of the image. Control adjustment is described in more detail in Chapter 10, and image manipulation using software is described in Chapter 11.

## 3.1 Display Types

The popular literature tends to describe displays as CRTs or flat-panel displays, but some CRTs have a flat face and flat-panel displays come in a variety of types. A CRT produces light by exciting a phosphor coating with an electron beam. Flat-panel displays are of two broad types[1]: one type filters reflected light or light from a source behind the filter, while the second type creates light by exciting a phosphor. Currently, image displays are largely of two types, cathode-ray tubes (CRTs) and active-matrix liquid crystal displays (AMLCDs). Plasma display panels (PDPs) also may soon be used for critical viewing. Other types of displays such as field-emissive displays (FEDs) and organic light-emitting diodes (OLEDs) are receiving considerable development effort but are not yet viable as image display alternatives.

The CRT has been the dominant soft-copy display technology for many years. Despite the proliferation of flat-panel displays, CRTs continue to be competitive in the marketplace.[2] They are manufactured as both monochrome and color displays and can be characterized as cheap and bulky. The term "monochrome" is commonly applied to gray-scale monitors, but technically they should be called achromatic monitors. This book will follow convention and use the term monochrome to refer to gray-scale monitors. A CRT requires a relatively large amount of power (more than an AMLCD but less than a PDP). A good-quality 17-in. diagonal CRT can be purchased for $150–$350 (U.S.), weighs 40–50 pounds, and occupies a space of

**Figure 3.1** CRT. Image courtesy of Siemens AG Display Technologies.

~1.7 cubic feet (Fig. 3.1). An AMLCD with the same viewing area costs $550–$1300, weighs about 20 pounds, and occupies ~0.6 cubic feet (Fig. 3.2). Until recently, the CRT has had more individually addressable pixels, although viewing quality may be comparable to an AMLCD. A pixel is the smallest display element in a digital or raster scan display; an addressable pixel is one that defines a unique location and can be commanded to a specific output luminance level. For an analog display, a pixel is related to the CRT spot size and scan lines. Recent developments have led to AMLCDs with very large numbers (up to 9.2 million) of addressable pixels;[3,4] AMLCDs are also being developed for high-definition TV (HDTV) applications.[5]

Plasma displays are currently being developed largely for HDTV viewing with 37-in. and larger screens.[5,6] They are expensive ($6000 and up), have roughly the same number of addressable pixels as 17-in. LCDs, can be hung on a wall like a picture, and offer a wide viewing angle with no loss in quality. They offer high brightness but have a relatively slow response time. Although plasma displays are not currently viable replacements for CRTs and LCDs, they could conceivably replace them in the future.

Displays can also be categorized in terms of monitor mode—portrait or landscape. Landscape monitors show an image that is wider than it is high, typically a ratio of 4:3 or 16:9. Portrait-mode monitors have reversed ratios and are more often used in the medical field.

**Figure 3.2** AMLCD. Image courtesy of Siemens AG Display Technologies.

## 3.2 Display Controller

Any electronic display requires a device that transforms the digital image into a signal that drives the display. These devices are called display controllers, video controllers, or simply graphics cards.[7] For CRTs, the digital image must be converted to an analog signal to drive the CRT electron gun or guns. AMLCDs and plasma displays can be driven with a digital signal, although many are currently driven with an analog signal or have the option of using either signal.

An analog signal constantly varies over a range of frequencies and amplitudes (Fig. 3.3).[8] Amplitude defines the magnitude of the signal, and frequency describes variation as a function of time. In order to convert an analog signal to a digital signal, the analog signal is sampled at a discrete frequency (e.g., $N$ samples per second). To accurately preserve the analog signal, the digital frequency is typically twice the highest frequency of the analog. The amplitude of the analog sample at each sample point is preserved as one of a discrete number of values. The total number of values is expressed as a power of 2. Thus, a system with $2^8$ discrete levels is termed an 8-bit system and has 256 discrete levels or values.

Analog TV cameras are analog devices in one dimension.[9] They record an analog signal in the horizontal dimension for each of a defined number of lines in the vertical (Fig. 3.4). They also sample in the time domain, providing a defined number of images (or lines) per unit time. In order to be recorded as a digital image, the

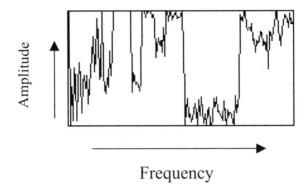

**Figure 3.3** Analog signal.

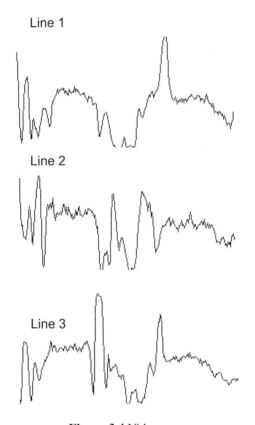

**Figure 3.4** Video scan.

analog signal must be converted to a digital signal. This process is called quantizing because a digital quantity is assigned to each amplitude value at each sample point (Fig. 3.5).

For an analog CRT, the digital image must be converted back to a continuously varying analog voltage signal or level. It is this voltage signal that begins the image

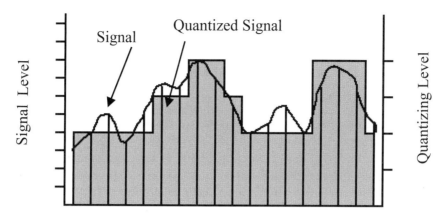

**Figure 3.5** Quantizing process.

display process. Since the most common displays are analog CRTs, computers typically output an analog signal to (or through) the display connector. This is true even for systems using digital displays, where there is no need for an analog signal. Attempts are being made to develop a digital standard for such displays that would eliminate the analog conversion.[10]

Aside from the basic analog versus digital difference, controllers differ in terms of a number of other parameters. They will be briefly discussed here, but because of their effect on the quality of the displayed image, controllers will be discussed in more detail in subsequent chapters.

Video controllers may operate at a fixed refresh rate, or at several. The refresh rate defines the frequency with which pixel values can be changed. A refresh rate of 75 Hz means that every addressable pixel is refreshed 75 times per second. The term "addressability" refers to the number of addressable pixels in the vertical and horizontal dimensions. Controllers may have a single allowable addressability value (e.g., 1600 × 1200) or the rates may be set by the user to any one of several allowable rates. Rates are sometimes also described by acronyms such as VGA (640 × 480) or XGA (1024 × 768). A CRT typically allows several rates, while an AMLCD typically has a single rate defined by the number of liquid crystal cells or addressable pixels. The refresh rate may be tied to addressability values, decreasing as addressability increases.

Some controllers provide a function that allows the display luminance values to be set to desired levels. Using an accompanying luminance measurement device and software, the user can set the display luminance values to the desired preprogrammed levels. Displays also exist that modify their luminance output to overcome, or attempt to overcome, the effects of room lighting falling on the face of the display.

Bit depth refers to the number of digital values—typically 8, but sometimes 10 or 12—that the controller can process. Some drivers may process at one level and output at a lower level. Color systems have three color channels. If each channel has an 8-bit capability, the system is said to be a 24-bit system. A 24-bit driver is capable of over 16 million colors, but memory limitations may reduce the actual number to as low as 256.

A display is refreshed many times per second. Most of the time, the same image values are displayed repeatedly. Rather than the image transmitting from the computer memory (CPU) to the display at that rate, two time-saving techniques are employed. The first is the use of a screen buffer, random access memory (RAM). The image is sent to the buffer once and the screen is refreshed from the buffer. Only when the image is changed does it become necessary for the image to be refreshed from the CPU. The second technique is called graphics acceleration. Many of the operations performed on an image, such as moving blocks of pixels, filling areas with the same digital values, and drawing lines (vectors), are simple and repetitive. When these operations are performed on the display controller, the graphics process can be accelerated.[11]

A variety of technical specifications relate to the interface between the controller and display (and connecting cable) and the controller and CPU. Proper integration is required or performance will suffer. It is assumed for purposes of this discussion that such integration has taken place. The key point to be remembered regarding display controllers is that they can significantly affect the final appearance of an image. Image appearance, including circumstances where the display controller may contribute to poor performance, will be discussed in greater detail in subsequent chapters.

## 3.3 CRT Operation—Monochrome

A monochrome CRT display is an analog device that operates by producing light energy from electrically stimulated luminescent phosphor. Phosphor is any material that emits light (luminesces) when struck by some form of radiant energy, in this case a current of electrons. A monochrome CRT has a single phosphor type that emits light of a single color or wavelength. Some early CRTs showed shades of green or orange. Current monochrome CRTs show shades of gray (black to white) and are thus achromatic. Figure 3.6 shows the basic operation of a monochrome CRT. The analog signal from the display controller is converted to voltage proportional to the signal amplitude. The voltage is then amplified in a nonlinear form. The form can be considered exponential such that

$$C = V^\gamma, \qquad (3.1)$$

where $C$ is the amplified beam current, $V$ is the amplified voltage, and $\gamma$ (gamma) is the exponent. In luminance space, the relationship can be defined as[12]

$$L = L_{min} + (L_{max} - L_{min})v^\gamma, \qquad (3.2)$$

where $v$ is the normalized input voltage. In theory, the relationship would be linear in log space and gamma would be a measure that described the relationship in log

# ELECTRONIC DISPLAY OPERATION

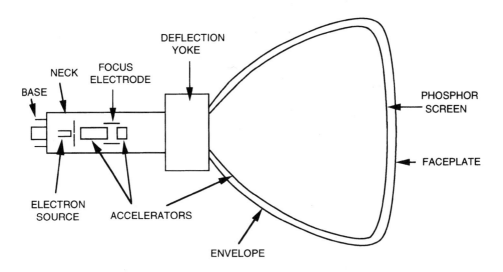

**Figure 3.6** Basic CRT operation. Reprinted by permission of Peter A. Keller from *Electronic Display Measurement* (John Wiley & Sons, Inc.). © 1997 Peter A. Keller.

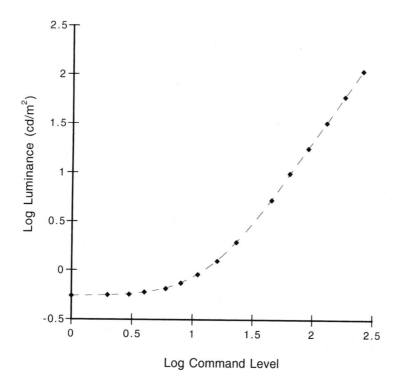

**Figure 3.7** Luminance vs. command level (CL).

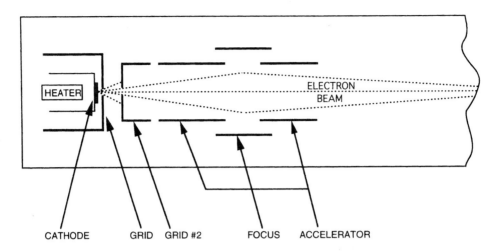

**Figure 3.8** Beam focusing. Reprinted by permission of Peter A. Keller from *Electronic Display Measurement* (John Wiley & Sons, Inc.). © 1997 Peter A. Keller.

space. Figure 3.7 shows a set of measurements on a CRT. The relationship is linear only above a certain command level (CL 16).

The process of producing a CRT image begins with the cathode. The cathode is heated so that electrons are emitted as a beam from the oxide coating. With this type of cathode, an oxide cathode, the oxide coating wears away over time and results in degradation of the light output. A second type of cathode, called the dispenser cathode, continuously replenishes the source of electrons and therefore offers greater longevity and better efficiency. A grid in front of the cathode controls the quantity of electrons in relationship to the original drive voltage, which in turn is proportional to the CL. The grid reduces the flow of electrons.

The electron beam is accelerated and focused to create increased luminance from the phosphor. Without focusing, the electrons would be dispersed all over the phosphor. The beam is focused with an electrostatic lens system (Fig. 3.8), which can correct for the defocusing that would otherwise occur in the corners of the display. This is called dynamic focusing.[13]

The components involved in generating and focusing the electron beam together form the electron gun. The gun shoots the electrons out to the phosphor screen. The stream of electrons is deflected in both the horizontal and vertical direction magnetically. This causes the beam to sweep across the display in the pattern shown in Fig. 3.9. Computer displays use progressive scan as opposed to the interlaced scan used in TV displays. With progressive scan, the image is scanned one line at a time from top to bottom and then the process is repeated to form the next image. With interlaced scan, every other line is traced on the first pass (field) and then the alternate lines are traced on the second field as shown in Fig. 3.10. Vision (and phosphor) persistence makes the image appear as a single frame. Interlacing allows the use of lower bandwidths.

# ELECTRONIC DISPLAY OPERATION

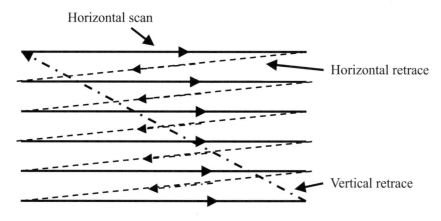

**Figure 3.9** Progressive scan pattern.

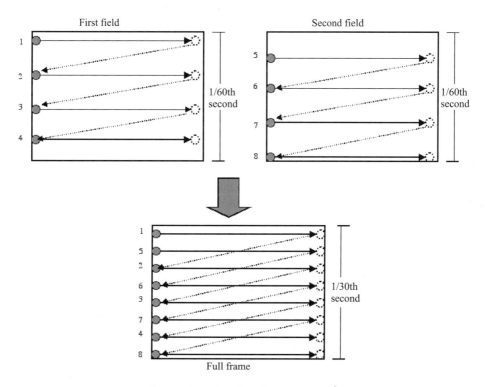

**Figure 3.10** Interlaced scan pattern.[9]

When the electron beam strikes the phosphor, light energy is emitted. The amount of energy emitted is a function of the electron beam energy and the phosphor. Phosphors differ in terms of persistence, luminance for a given drive level, and color. Longer-persistence phosphors can result in a more stable image but at the expense of blurring when a fast refresh is required. Higher luminance at the

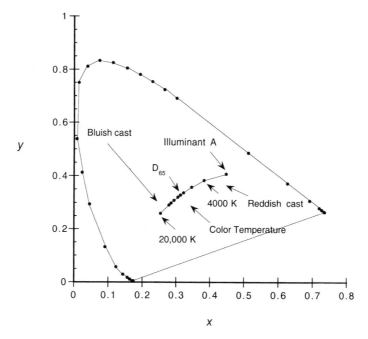

**Figure 3.11** White color temperature. Data from Ref. [14] based on CIE 1931 coordinates.

same beam energy results in greater efficiency. Although monochrome monitors are in theory achromatic, they may exhibit a slight color cast because of the characteristics of the phosphor. The color cast is described in terms of color coordinates. For a nominal white phosphor, the color coordinates indicate whether the white has a blue or red cast (Fig. 3.11).

The light energy emitted by the phosphor passes through a face plate to the observer. The face plate is often treated to achieve one or more desired goals, such as reducing the effect of room light striking and reflecting from the screen (antireflection coating or etching). For this goal, the treatment might involve applying a filter to reduce the room light passing through the face plate and adding to the light energy from the phosphor. Similarly, glass with different transmission properties may be used. However, although the use of reduced transmission glass or filters increases contrast, it also either reduces the light emitted by the phosphor or requires a higher level of electron energy to maintain luminance. As the energy is increased, it becomes more difficult to maintain a focused point of luminance; thus, quality may be degraded. Since more electrons may be needed, longevity of the cathode dispenser may also suffer.

The final relationship between the original image digital values and the emitted light energy can also be described by a curvilinear function of the type

$$L = a + bCL^c, \tag{3.3}$$

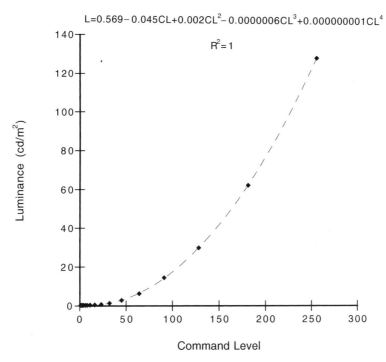

**Figure 3.12** CRT command level/luminance function.

where $L$ is output luminance, $CL$ is the input command level, and $a, b,$ and $c$ are coefficients. Figure 3.12 provides an example. The function can be approximated with a fourth-order polynomial as shown in the figure.

The elements of the CRT are enclosed in a glass tube, which maintains a vacuum to increase the efficiency of the electron beam. The tube must be of a shape and construction such that it can resist the stress resulting from atmospheric pressure on the vacuum within the tube. It is for this reason that many CRTs have a curved surface. Unfortunately the curved shape introduces distortion, and the size and shape of the tube and gun are bulky. Tube depth is about the same dimension as the tube face plate's diagonal dimension. Perhaps prompted in part by the competition from non-CRT flat-panel displays (LCDs and plasma), two types of so-called flat-panel CRTs have emerged. The first is a true flat panel that substantially reduces the depth of a normal CRT. The cathode is either placed below the phosphor screen with one or more 90-deg bends in the beam or is an area cathode with aperture grids. Such devices have not yet achieved commercial success.[13] Recently, the process of manufacturing CRT tubes has evolved to create the second type of flat-panel CRT, in which the curvature of the face plate is reduced or eliminated. The depth remains approximately the same as conventional CRTs. Reduced-depth CRTs are also being produced by increasing the allowable deflection angle. The reduction is relatively small (about 2 in. for a 17-in. diagonal monitor).[15]

## 3.4 CRT Operation—Color

The operation of a color CRT is similar to that of a monochrome CRT except that three electron beams and three phosphors are required to produce color. As mentioned in Chapter 2, electronic display color is called additive because three colors—red, green, and blue—are added together to produce a wide variety of colors. Each of the three colors has a beam that illuminates a phosphor that emits red, green, or blue light. The phosphor is laid down as alternating color stripes on the screen or as individual dots. Each pixel is made up of three separately illuminated phosphor dots or lines that visually combine to produce a single color. In order to ensure that each beam strikes its associated phosphor, a mask is generally employed. The mask, sometimes called a shadow mask, can be designed for a triangular arrangement of phosphor dots or for a vertical arrangement (Fig. 3.13). For vertically arranged dots, the mask may be a vertical grill or a vertically slotted mask.[16] For a variety of reasons, the beam and mask or grill apertures may not align with the three color beams, resulting in a loss of color purity because the beams and their respective phosphors are no longer in alignment.

Because color CRTs require three beams and the mask and grill techniques, color displays tend to show both reduced resolution and a loss of energy efficiency relative to monochrome displays. Resolution is diminished by limits on the ability to focus and control the position of the electron beam since it is easier to control a single beam than three beams. Also, the energy that strikes the grill or mask is lost and may heat the grill or mask, resulting in further distortion.

Theoretical alternatives to the use of a mask or grill have been proposed, including sequential color separation (SCS)[17] and "fast intelligent tracking."[18] With SCS, the three color beams are switched at a rate proportional to the scan fre-

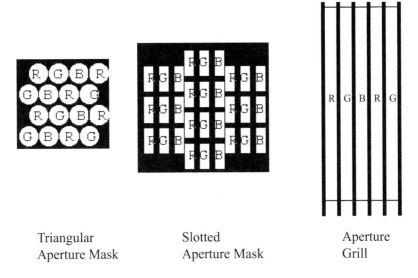

Triangular Aperture Mask   Slotted Aperture Mask   Aperture Grill

**Figure 3.13** Color CRT aperture types.

quency such that the proper beam strikes sequentially laid color phosphors. Although early attempts were not successful, a similar technique called dynamic color separation (DCS) has recently been proposed. Whether it is a viable alternative to current techniques remains to be proven. In the case of the fast intelligent tracking tube, beams are scanned along horizontal phosphor lines with position correction provided by a conducting detector above and below each phosphor line. Proof of this principle has been demonstrated.[18]

## 3.5 The AMLCD

The AMLCD (Fig. 3.14) is currently the leading flat-panel display. It is termed "active" because it uses a transistor to drive the state of each pixel and contains a matrix of liquid crystals (and related control). A passive-matrix LCD uses an array of row and column wires to change the display in a scanning fashion. Passive-matrix displays have low contrast ratios and slow response times compared with AMLCDs.

Liquid crystals are oblong organic molecules whose alignment can be changed with an electrical field. Figure 3.14 shows a typical alignment pattern for a twisted nematic (TN) LCD. The molecules are placed between two grooved glass plates with the grooves at 90 deg to each other. This causes the molecule alignment to spiral between the two plates. Polarizers located above and below the glass plates are used to control light transmission.[13] Incoming light is polarized by placing a polarizer between the light source and the liquid crystal. As light passes through the plates and crystals, its polarization may be changed. For instance, when an electrical field is applied (on state), the molecules align with the field and the light is polarized. If the molecules are rotated such that they are at 90 deg to the output plate grooves (and polarizer), the light is reduced to a minimum. If they are parallel to the output polarizer, light output is at a maximum. In an AMLCD, each pixel has

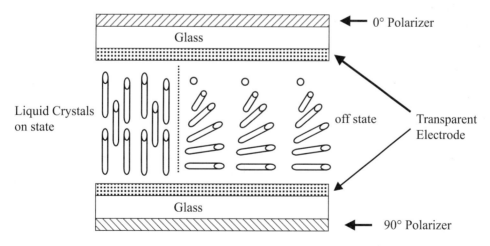

**Figure 3.14** AMLCD operation.

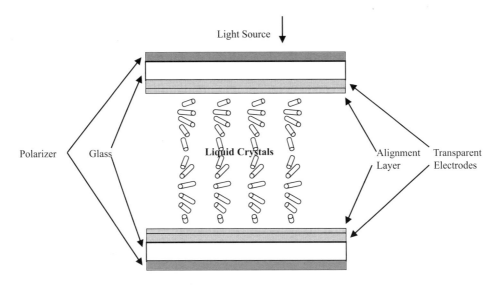

**Figure 3.15** STN operation.

an associated transistor that controls the voltage and thus the alignment of the molecules. The use of individually addressable transistors speeds up the process of creating an image. By using three filters (red, green, and blue), color can be obtained. Each pixel is then made up of three subpixels, each controlled by its own transistor.

The use of polarization as a technique to control light intensity has the disadvantage of reducing acceptable viewing angle. As one moves off-axis (relative to the polarization axis), the intensity changes so contrast reversals and color changes can occur. The use of so-called supertwisted nematic (STN) LCDs has alleviated this effect (Fig. 3.15);[13] STNs show 270 deg of rotation and have a sharper transition between on- and off-transmittance conditions.

A second quality issue is that of noncommandable pixels. Manufacturing defects can result in nonoperable cells or pixels that remain in an on or off position regardless of the CL sent to them.

The input/output (I/O) function of an AMLCD (Fig. 3.16) is typically somewhat less linear in log/log space than is the case with a CRT. With a digital driver, the function can be anything the user desires.

The AMLCDs currently on the market typically show 1600 × 1200 addressability for a 22-in. diagonal display. The viewable area of a 22-in. AMLCD is equivalent to a 24-in. CRT. The maximum luminance exceeds that of a color CRT. The power required is about half that of a CRT, and as of early 2002, the list price was about twice that of a 24-in. CRT. The addressability of AMLCDs continues to increase: One monitor currently on the market shows 3840 × 2400 pixels.[4]

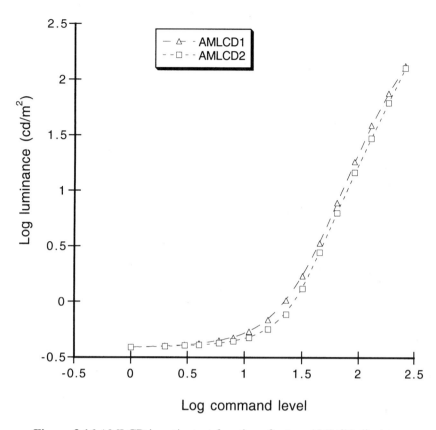

**Figure 3.16** AMLCD input/output functions for two AMLCD displays.

## 3.6 Plasma Displays

Plasma display panels (PDPs) developed from early neon glow lamps.[13] A tube inside the lamp was filled with neon that glowed when the gas was ionized by the application of a current. Arranged in rows, the lamps could be used to represent numbers. Current PDPs employ phosphor coatings that are excited by discharge from the ionized gas contained between two layers of glass. Figure 3.17 shows one of the currently used designs (alternating-current, matrix sustain structure, or ACM). A pixel is defined by three phosphors and a matrix of row and column electrodes. Ribs separate the column electrodes. The space between the glass is filled with a mixture of neon and xenon. When a current is passed through the gas, it emits light in the UV portion of the spectrum. This excites the colored phosphors that emit visible light energy. The energy from the three phosphors mixes to produce color. Once the gas has begun to emit light, it continues to do so until the device is turned off. The cell thus has a built-in memory effect. Unlike the case with CRTs, the relationship between digital image value and output luminance is linear. This creates more of a mismatch between the display and the response of the HVS. Chapter 6 will discuss this mismatch in detail.

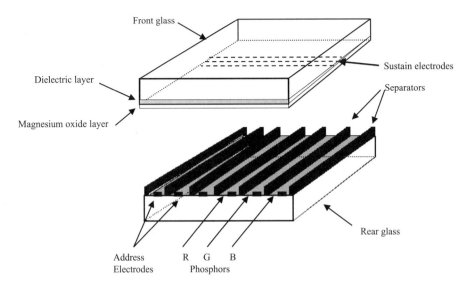

**Figure 3.17** An ACM plasma display panel operation.

Plasma display panels are currently of great interest for HDTV applications. Cathode-ray tubes are not practical for large (40- to 50-in.) TV displays because of their weight and size. A PDP offers high brightness and good viewing angle. As is the case with AMLCDs, luminance (and color) and pixel shape are uniform across the display. Contrast can be limited, and because of slow refresh rates, motion can be a problem. Like AMLCDs, cells (pixels) can be inoperable. Plasma display panels currently on the market are almost uniformly large (36 to 60 in.) and costly ($6000 U.S. and up). It remains to be seen whether or not PDPs will be competitive in the image display market.

## 3.7 Display Controls

Depending on the type of display, three or four types of hardware controls are available to modify the appearance of images on a display. All types of displays have luminance controls, typically contrast and brightness. Cathode-ray tubes have geometric distortion controls that are used to align the image on the display and may also have a display frequency control (refresh rate) and an addressability control. Color monitors have color controls that may include color temperature, color bit depth, and color gain.

### 3.7.1 Luminance controls

Most displays have brightness and contrast controls. Brightness increases the absolute luminance levels of the display. The luminance of both dark and light areas is increased, but not necessarily by the same amount. The difference between the

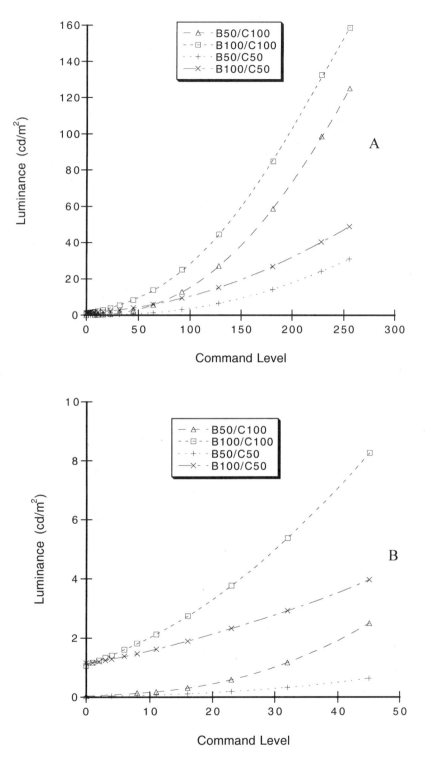

**Figure 3.18** Effect of brightness and contrast controls.

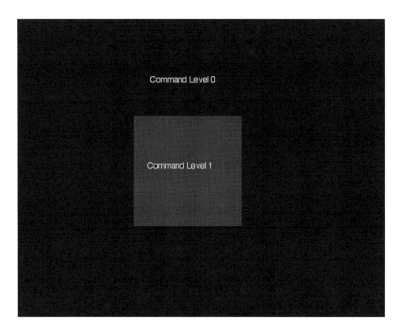

**Figure 3.19** Dark cutoff target.

lowest and highest brightness areas may also change. Contrast increases the difference between the darkest and lightest luminance areas, generally by increasing the luminance of the brightest area. Some AMLCDs also have the ability to vary the background light level (the light source that is then reduced in intensity by the polarizing effect of the TN or STN cell).

The effect on image appearance of varying brightness and contrast is not always obvious. Figure 3.18 shows the effect of varying contrast (C) and brightness (B) on a CRT in terms of measured luminance values. Figure 3.18(A) shows the full CL/luminance range, while Fig. 3.18(B) shows the lower portion of the range for greater clarity. Note that changes in both brightness and contrast affect the maximum output luminance, and this relationship is not intuitive. Generally, it is best to set brightness at the lowest level that avoids dark cutoff, which occurs when changes in low command values do not produce a change in output luminance. The target shown in Fig. 3.19 can be used to test for the effect. The center dark square (CL 1) should be discernable in the zero CL background. The use of contrast and brightness controls is discussed further in Chapter 10.

### 3.7.2 Geometry controls

Because the electron beam on a CRT is magnetically deflected, various controls to ensure linearity are typically provided. Figure 3.20 illustrates different types of distortion that may be present. The pattern shown as Fig. 3.21 may be used to adjust the various controls.

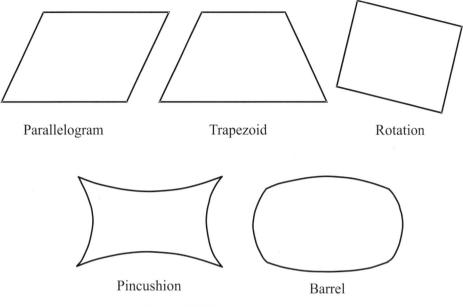

**Figure 3.20** Geometric distortions.

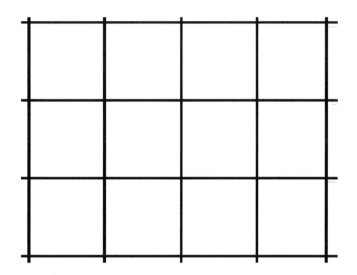

**Figure 3.21** Distortion target.

AMLCD and PDP geometries are defined by the individual grids or cells. No adjustment for linearity is thus needed or generally possible. However, position controls may be available to move the image vertically or horizontally. In addition, AMLCDs may have a noise adjustment control to reduce apparent noise in the displayed image.

Displays may have adjustable addressabilities and refresh rates. Addressability defines the number of uniquely addressable pixels; the refresh rate defines the

640 × 480 pixel addressability

1024 × 768 addressability

**Figure 3.22** Addressability comparison.

number of times the display is refreshed each second. The range of addressabilities may be constrained by the range of refresh rates. The refresh rate should generally be maintained at 75 Hz or greater, or undesirable flicker can occur. The display will usually appear sharpest when addressability is set to the highest level, but the finest detail displayed on the monitor may not be discernable (see Chapter 6). Figure 3.22 shows an image at two different addressabilities; at the lower addressability, the pattern of scan lines, called the raster pattern, is evident.

### *3.7.3 Color controls*

Three color controls are typically available on color displays. The first is color temperature, although it may not always be indicated as such. Color temperature refers to the color coordinates of the monitor white point (see Fig. 3.11). The monitor white point is defined by the color coordinates when all three color guns are at their maximum luminance. Color temperatures are defined (in Kelvin units) by the temperatures of colors emitted by a blackbody (a perfect emitter and absorber), which always emits the same color at any given temperature. Thus, a color temperature of 5600 K means that the white point has the same color as a blackbody heated to a temperature of 5600 K. As temperature increases, the white changes from a red to a blue cast. The maximum achievable luminance also increases as color temperature increases. Monitors typically show a range between 5000 K and 9300 K.

A second control that may be present is color gain or adjustment. This is sometimes called user-adjustable temperature or white point. On a CRT, color is produced by a mixture from the red, green, and blue guns exciting the red, green, and blue phosphors (or, for a PDP, by UV energy exciting the RGB phosphors). If all of the guns are off, black is perceived. If all are on at full strength, white is perceived. If the strength varies, the white will have some color cast depending on the relative distribution. In general, setting the gains to the maximum will maximize luminance.

The final color control is typically related to the display controller. The controller sets the number of displayed colors. A typical display has 8 bits per color channel and thus can display over 16 million colors. For various reasons, it is sometimes considered desirable to reduce the number of displayed colors to thousands or even as few as 256. This is sometimes called pseudocolor. Unless demanded by a software application, there is usually no reason to adjust the control for the number of colors when displaying color imagery.

## 3.8 Summary

Cathode-ray tubes and AMLCDs are the dominant display technologies used for image display. PDPs currently seem to be designed more for the HDTV market. Other display types such as FEDs and OLEDs have not yet been developed to the point where they produce viable image displays.

Display controllers or graphics cards translate the digital command values of an image into an analog or digital electrical signal. In the case of a CRT, the signal drives an electron dispenser. The electrons are accelerated, focused, and directed at a phosphor coating on the rear of the display face. The phosphor emits light energy through the display face plate to form a luminance image. In the case of a color display, three electron beams are used, one for each color (R,G,B). Some type of mask or grill directs the energy to a specific color phosphor; by varying the energy directed to each color phosphor, a wide variety of colors can be produced.

For an AMLCD, the polarization of a liquid crystal (for each pixel) is controlled by a transistor as a function of the digital command value. The variation in polarization reduces the light transmitted by a display light source. Separate crystals with color filters are used to produce color.

Plasma display panels use ionized gas to excite phosphors. The gas is a mixture of neon and xenon and emits light in the UV portion of the spectrum; the phosphors emit light in the visible portion of the spectrum. Once the gas begins to emit light, it continues to do so until turned off.

Most displays have controls for luminance (brightness and contrast), color (temperature, color gain, quantization), and geometry (distortion correction). Geometric distortion can generally be eliminated through use of the controls. Brightness should be set so as to eliminate dark cutoff, and the brightness and contrast controls together should be set to produce values as recommended in Chapter 10. Chapter 10 also discusses optimization of color control settings.

## References

[1] H. F. Gray, " The field emitter display," *Information Display*, Vol. 3, pp. 9–14 (1993).
[2] M. Maeda, "Will CRTs survive?" *Information Display*, Vol. 3, pp. 24–27 (1999).
[3] H. Blume et al., "Characterization of physical image quality of a 3-million-pixel monochrome AMLCD: DICOM conformance, spatial MTF, and spatial NPS," Society for Information Display, 2002 International Symposium, Digest of Technical Papers, Vol. XXXIII, No. 1, pp. 86–89 (2002).
[4] Y. Hosoya and S. L. Wright, " High-resolution LCD technologies for the IBM T220/T221 monitor," Society for Information Display, 2002 International Symposium, Digest of Technical Papers, Vol. XXXIII, No. 1, pp. 83–85 (2002).
[5] J. H. Souk and N. D. Kim, "40 in. wide XGA-TFT-LCD for HDTV application," Society for Information Display, 2002 International Symposium, Digest of Technical Papers, Vol. XXXIII, No. 2, pp. 1277–1279 (2002).
[6] Y. Sato et al., "A 50 in. diagonal plasma display with high luminous efficiency and high display quality," Society for Information Display, 2002 International Symposium, Digest of Technical Papers, Vol. XXXIII, No. 2, pp. 1060–1063 (2002).
[7] H. Roehrig, "The monochrome cathode ray tube display and its performance," in *Handbook of Medical Imaging*, Y. Kim and S. Horii (eds.), Vol. 3, Displays and PACS, SPIE Press, Bellingham, WA, pp. 157–220 (2000).
[8] A. F. Inglis and C. Lutha, *Video Engineering*, Second Edition, McGraw Hill, New York (1996).
[9] R. J. Farrell and J. M. Booth, *Design Handbook for Imagery Interpretation Equipment*, Boeing Aerospace Co., Seattle, WA (1984).
[10] C. Jacobson, "Digital visual interface comes into focus," *Intel Developer Update Magazine*, Issue 22, http://intel.com/update/archive/issue/stories/top6.htm (July 2001).
[11] F. M. Behlen, B. M. Hemminger, and S. C. Horii, "Displays," in *Handbook of Medical Imaging*, Y. Kim and S. Horii (eds.), Vol. 3, Displays and PACS, SPIE Press, Bellingham, WA, pp. 405–440 (2002).
[12] E. Muka, H. Blume, and S. Daly, "Display of medical images on CRT softcopy displays: a tutorial," *Proc. SPIE*, Vol. 2431, pp. 341–358 (1995).

[13] P. A. Keller, *Electronic Display Measurement*, John Wiley & Sons, Inc., New York (1997).

[14] G. Wyszecki and W. S. Stiles, *Color Science*, Second Edition, John Wiley & Sons, Inc., New York (2000).

[15] A. Kato et al., "Development of a 41-cm (17-in.) short-length color monitor tube," Society for Information Display, International Symposium, Digest of Technical Papers, Vol. XXVIII, pp. 138–141 (1997).

[16] G. Spekowius, M. Weibrecht, P. Quadfleig, and H. Reiter, "Image quality assessment of color monitors for medical softcopy display," *Proc. SPIE*, Vol. 3658, pp. 280–290 (1999).

[17] C. Washburn, "End of the shadow mask?" *Information Display*, Vol. 6, pp. 16–20 (1998).

[18] A. H. Bergman et al., "The fast intelligent tracking (F!T) tube: a CRT without a shadow mask," Society for Information Display, 2002 International Symposium, Digest of Technical Papers, Vol. XXXIII, No. 2, pp. 1210–1213 (2002).

# Chapter 4
# Physical Display Quality Measures

The previous chapter described the operation of displays and discussed the effect of hardware controls on image appearance. This chapter describes how physical quality measures are used to characterize the performance of a display. Chapter 5 will describe perceptual measures of quality and utility. Previously quality was defined as the degree to which a displayed image accurately portrays the original "scene" (e.g., a lung or a section of terrain) and accuracy was defined in terms of spatial and spectral response. Quality can thus be defined in terms of a mathematical relationship between the image and scene and is a function of both the image capture device and the display. Quality can be measured in both the physical and perceptual domains. The true scene/image relationship is defined in the physical domain, while perceptual judgements can be used to estimate the relationship. For example, contrast can be defined in terms of luminance differences or in terms of perceived brightness differences. Utility, on the other hand, is the value of the displayed image to an observer. Value is typically measured in terms of the ability to perform a specified task. Whereas quality can be measured in both the physical and perceptual domains, utility is measured in the perceptual and cognitive (eye and brain) domains.

Many measures have been devised to characterize and report on the quality of displays. This chapter describes the measures used in later chapters to select and optimize displays, although not all of the measures described here will be used in this book. They are discussed because readers may encounter them in other display literature. In the following chapters, we use a subset of these measures to define desired display performance.

Physical measures of display quality can be categorized in terms of resolution, contrast, noise, and artifacts and distortions. Resolution relates to the ability to see fine detail in an image. Contrast refers to tonal or color differences within the image. Noise refers to unwanted signal variation, either spatially or over time. Artifacts are unwanted image variations generally resulting from processing or signal transmission. Distortions are defined in terms of departure from straightness and linearity. Physical measures can also be categorized in terms of the measurement domain. Measurement domains include spatial, luminance, spectral (color), and temporal or time-related.

## 4.1 Resolution Measures

The ability of a monitor to portray image detail is a function of pixel size and shape. It is also a function of contrast. Thus, although we separate resolution and contrast measures in this discussion, they act together to determine our ability to see image detail.

### 4.1.1 Addressability and screen size

The pixel represents the smallest potential level of image information that can be seen on a display. However, on a color monitor, a pixel is made up of three subpixels, each defining a different color. The number of pixels on a monitor is defined by its addressability, which is the number of uniquely commandable picture elements (pixels). Addressability is usually expressed in the vertical and horizontal dimensions. Thus, a monitor with 1600 × 1200 addressability has 1600 pixels in the horizontal dimension and 1200 in the vertical.[1]

Addressability indicates the number of pixels on a monitor but not the size of the pixels. To determine pixel size, the user must know the size of the monitor display area. Monitors are commonly defined in terms of the dimension of the diagonal (Fig. 4.1). Most monitor descriptions specify the size of the viewable area as

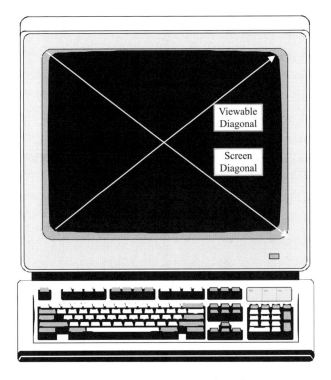

**Figure 4.1** Monitor size designation.

less than the full screen. Until the advent of HDTV, the monitor ratio of width to height was 4:3; HDTV format is 16:9. These ratios apply to a landscape format (width greater than height). They are reversed for a portrait-format monitor, where height is greater than width. Portrait-format monitors are commonly used in the medical profession because they more nearly represent the format of x rays. The portrait format has also been shown to be preferred for reading text.[2]

If the diagonal dimension of a monitor is known, the vertical and horizontal dimensions can be computed for a 4:3 landscape display by

$$H = 3\sqrt{\frac{D^2}{25}} \text{ and}$$
$$W = 4\sqrt{\frac{D^2}{25}}.$$

(4.1)

For a 16:9 landscape display, they are computed by

$$H = 9\sqrt{\frac{D^2}{337}} \text{ and}$$
$$W = 16\sqrt{\frac{D^2}{337}}.$$

(4.2)

The dimensions are simply reversed for the portrait format.

### 4.1.2 Pixel density and size

Pixel density is the number of pixels per unit dimension, often expressed as pixels per in. (ppi). If the width and height of a display are known, pixel density can be computed. For example, a 4:3 display with an 18-in. diagonal has a display width of 14.4 in. and a height of 10.8 in. If the addressability is 1600 × 1200, pixel density is 111 ppi. To arrive at this figure, simply divide the number of pixels by the display dimension (i.e., 1600 ÷ 14.4 = 111.11). The vertical dimension will yield the same result (1200 ÷ 10.8 = 111.11).

Pixel density can be used to define the spacing of pixels, called pixel pitch. If we have 111 ppi, then each pixel is spaced 0.009 in. from the next pixel. Pixel pitch, $P$, is computed as

$$P = \frac{1}{PD},$$

(4.3)

where $PD$ is pixel density. For reasons unknown to the author, screen dimensions are typically defined in inches and pixel pitch in millimeters. The conversion is

$$1 \text{ in.} = 25.4 \text{ mm and} \qquad (4.4)$$
$$1 \text{ mm} = 0.0937 \text{ in.}$$

For monochrome CRT and AMLCD displays, the concept of pixel pitch is straightforward since a pixel is defined by a single phosphor location. With color CRT and plasma displays, a pixel is defined by three phosphors. For shadow mask displays, the pixel pitch distance is the diagonal distance between like-color phosphors (Fig. 4.2). For aperture grill and plasma displays, pitch is measured in one dimension as the distance between phosphors of the same color (e.g., green to green).

For a CRT, a pixel is represented by a nonuniform energy distribution. The pixel width or size is normally defined as the width at the half-power level as shown in Fig. 4.3.[3] The pixel shape is defined by the relationship of the pixel width or size measured in both the vertical and horizontal dimensions. With color displays, a pixel is represented by three energy distributions. Line width or pixel size is defined in terms of a Gaussian curve fitted to the three energy response functions.

Spot or pixel size typically varies as a function of intensity. It is thus important to measure pixel size at the maximum intensity at which the monitor will be operated. Conversely, one can measure pixel size at various intensities to define the maximum useful intensity (the point beyond which resolution degrades because of the pixel size increase).

If the electron beams on a color monitor are misaligned, the three beams will not properly converge and a color fringe or edge (and a loss of resolution) will result. Convergence (or misconvergence) measures the degree of misalignment in

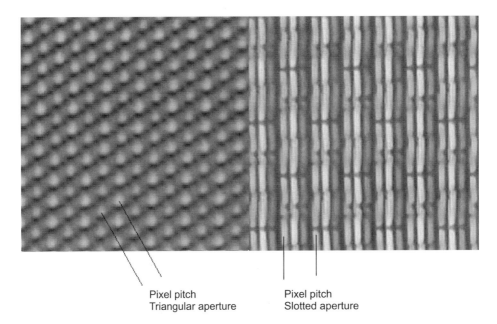

Pixel pitch
Triangular aperture

Pixel pitch
Slotted aperture

**Figure 4.2** Pixel pitch measurement.

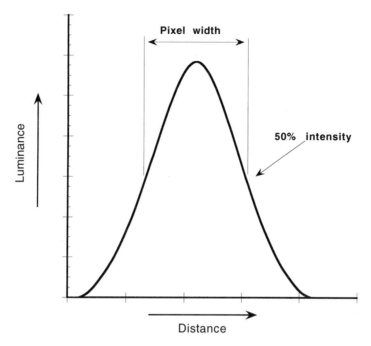

**Figure 4.3** Pixel size measurement.

terms of the misconvergence of blue and green with respect to red in either screen dimensions or fractions of a pixel size. [4]

A measure called "pixel fill factor" is used for AMLCDs.[5] Pixel fill factor measures the percentage of the pixel area that provides useful luminance.

## 4.1.3 Pixel subtense

We can also define some useful measures of resolution in terms of the observer. Observer performance is measured with the visual angle, the angle subtended at the eye (Fig. 4.4). If we define a viewing distance (*VD*) as the distance between the display and the observer's eyes, we can define the visual angle subtended by a pixel (the pixel subtense or *PS*) as

$$PS = \arctan\left(\frac{P}{VD}\right). \quad (4.5)$$

For a pixel pitch of 0.009 in. and a viewing distance of 18 in., the pixel subtense is 0.02865 deg (1.71 minutes).

Visual measures of resolution are often expressed in cycles per degree of visual angle subtense. A cycle is defined as an on-off sequence (Fig. 4.5). For the example previously given, one cycle subtends 0.0573 deg or 3.42 minutes. Thus, there are 17.45 cycles in a degree (17.45 cy/deg).

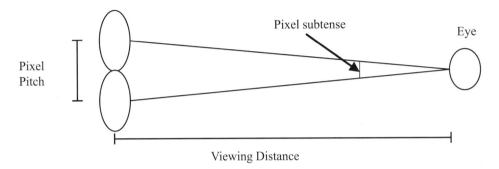

**Figure 4.4** Visual angle measurement.

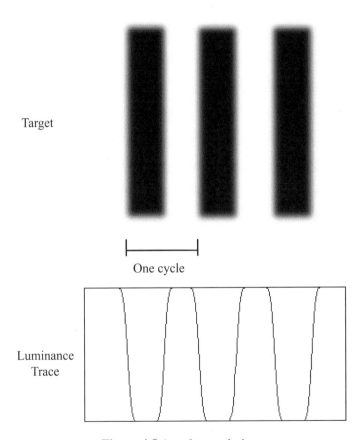

**Figure 4.5** Angular resolution.

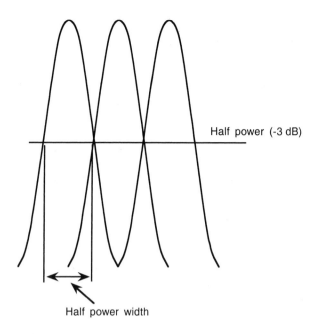

**Figure 4.6** RAR measurement.

## 4.1.4 Resolution-addressability ratio

Although a pixel is the smallest image element of a display, it is not necessarily perceived as such. In order to avoid the appearance of lines on a CRT display resulting from gaps between pixels, the energy distribution from adjacent pixels may be allowed to overlap. The degree of overlap is specified by the resolution-addressability ratio, RAR,[6] which is defined as

$$\text{RAR} = \frac{W_{-3dB}}{P}, \tag{4.6}$$

where $W_{-3dB}$ is the half-power width of the pixel point spread function and $P$ is the pixel pitch. Figure 4.6 depicts the concept. The RAR is typically in the range of 1.0 to 1.3 for CRT displays; AMLCDs and PDPs also show overlapping energy distributions.

## 4.1.5 Edge sharpness

Although not a direct measure of resolution, the sharpness of edges affects the perception of image and display quality.[7,8] Pixel density and subtense define the limits on visual details but do not define the sharpness of the displayed image. A measure called the relative edge response (RER) has been used to characterize edge sharpness for imagery. The RER is defined as the slope of the normalized edge response and is computed based on the system's modulation transfer function (MTF).[9] Figure 4.7 shows a normalized edge response. The RER is computed as

$$\text{RER}_x = ER_x(0.5) - ER_x(-0.5) \text{ and} \tag{4.7}$$
$$\text{RER}_y = ER_y(0.5) - ER_y(-0.5),$$

where ER is the slope of the edge response over the indicated region. The geometric mean (GM) is used as a single value of RER and is computed as

$$\text{RER}_{GM} = \sqrt{RER_x * RER_y}. \tag{4.8}$$

The RER can be measured on the display or can be computed from the display's MTF.

A measure related to RER called edge response is sometimes used to describe a display. The edge response is simply the number of pixels required to transition from maximum to minimum luminance across a knife edge.[10] Figure 4.8 shows an example that is an enlargement of a screen shot; the actual edge was two pixels wide. Note that this definition of edge response is different than that referred to in Eq. (4.7).

### 4.1.6 Contrast modulation

A theoretically perfect display would show complete independence among pixels such that the luminance of a pixel would be affected only by its commanded digital

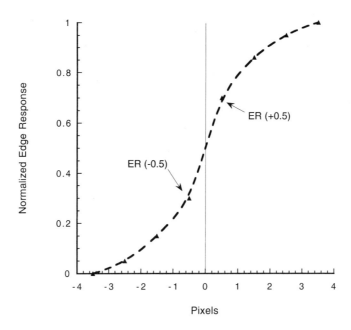

**Figure 4.7** RER measurement. Reprinted with permission from *Surveillance and Reconnaissance Imaging Systems: Modeling and Performance Prediction*, by Jon C. Leachtenauer and Ronald G. Driggers, Artech House Inc., Norwood, MA, USA (www.artechhouse.com).

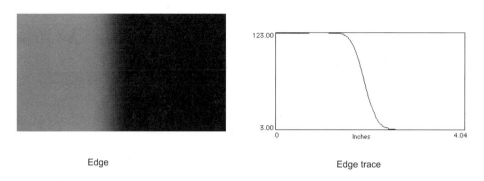

**Figure 4.8** Edge response.

**Figure 4.9** Cm measurement.

value or CL. The edge response would thus be 1. However, this is not the case because the luminance distribution of a pixel is affected not only by its CL but by those of surrounding pixels. One method of describing the effect is through the use of a measure called contrast modulation (Cm),[1] which is defined as

$$\mathrm{Cm} = \frac{L_B - L_D}{L_B + L_D}, \qquad (4.9)$$

where $L_B$ is the luminance of a bright line or patch and $L_D$ is the luminance of an adjacent dark line or patch. Figure 4.9 shows the method typically used to define Cm. Horizontal and vertical grills commanded to the maximum and minimum CL are measured. As the grill spacing gets smaller, Cm decreases (as shown in Fig. 4.10).

Summary measures of Cm performance are sometimes used. Cm uniformity measures the variation in Cm across the face of the monitor. A small variation is better than a large variation. A measure called "resolvable pixels" defines the number of pixels exceeding a defined Cm threshold (25% for imagery and 50% for text).[5] If the one on/one off measurements exceed the target value (e.g., 25%), the number of resolvable pixels equals the number of addressable pixels. If the target value is not reached with the one on/one off pattern but is reached with a coarser grill (two or three on/off), linear interpolation is used to define the noninteger spacing at which the target value is met (e.g., 1.25 pixels). The addressability value is divided by the interpolated pixel value to define the number of resolvable pixels. The worst

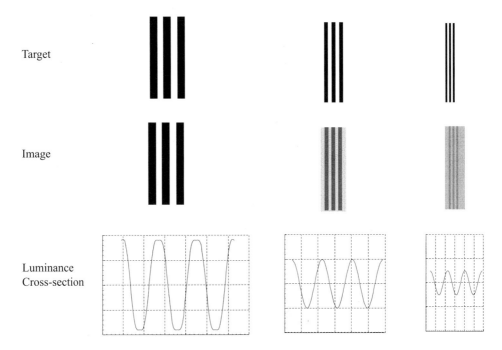

**Figure 4.10** Cm vs. grill spacing. Reprinted with permission from *Surveillance and Reconnaissance Imaging Systems: Modeling and Performance Prediction*, by Jon C. Leachtenauer and Ronald G. Driggers, Artech House Inc., Norwood, MA, USA (www.artechhouse.com).

case Cm measurement is used where "worst case" is defined by the square root of the sum of the squares of the vertical and horizontal Cm measures. Once the worst measurement point has been found, the vertical and horizontal measures are treated separately. Both values are reported just as they are for addressability.

Contrast modulation performance is not constant across the face of a CRT, and perceived Cm is not constant across an LCD monitor. It is thus useful to define performance in both a center area and a peripheral area of the monitor. The center area may be defined in terms of a portion of the total area (e.g., center 40%) or in terms of a size (e.g., 8-in. circle) based on a defined field of view at an assumed viewing distance.[11,12]

### 4.1.7 Raster modulation

A measure related to both Cm and RAR is called raster modulation.[13] Raster modulation is the variation in flat-field luminance resulting from the variation in energy across each pixel. It is measured using Eq. (4.9) with a vertical scan across a flat field, typically at the maximum luminance (Lmax) level of the display.

## 4.1.8 Modulation transfer function

Closely related to Cm is the MTF of the display. Whereas resolution is a single number defining the ability to see detail at a particular frequency (or the frequency at which detail can no longer be seen), the MTF defines the ability to see contrast as a function of frequency. The Cm measurements made at the various line spacings represent modulation values at 0.5 (one on/one off), 0.25 (two), and 0.167 (three) cycles per pixel. Figure 4.11 shows a plot for both the vertical and horizontal dimensions of a color CRT.

If the value of RAR is known and a Gaussian spot shape is assumed, the MTF of the display can be calculated. Contrast modulation values are computed using

$$Cm = \frac{2}{\pi}\exp\left[3.6\left(\frac{RAR}{2}\right) - \left(\frac{RAR}{2}\right)^2 + \left(\frac{RAR}{2}\right)^3\right]. \qquad (4.10)$$

The equation is valid over the range of 0.175<RAR<1.2.[6] For a display with an RAR of 1.2, the Cm value for the one-on/one-off condition (0.5 cy/pixel) is 0.551.

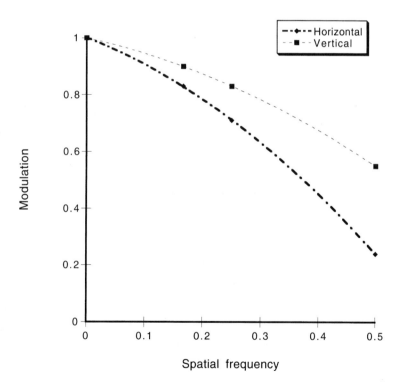

**Figure 4.11** Display MTF.

## 4.1.9 Bandwidth

A final measure sometimes used to characterize resolution is bandwidth. With an analog system, signal attenuation occurs as frequency increases. At some point, signal variations are no longer detectable. Bandwidth is a measure of the range of frequencies where signals are detectable above noise.[14] Specifically, it is the highest frequency transmitted. It is a measure that has been used with analog TV systems to define the effective horizontal resolution. The highest useful frequency relates to the minimum detectable horizontal spatial detail. It is sometimes specified for a monitor in terms of horizontal scan rate defined in kHz (thousands of cycles per second). In terms of resolution, the larger the bandwidth, the better.

## 4.2 Contrast Measures

In addition to Cm, several other measures are used to describe the ability to discriminate points or areas on a display that are commanded to different values (tone or color). Contrast measures are also called photometric measures. In addition to the standard measures used with CRTs, AMLCDs have a set of measures related to the effects of the viewing angle on luminance output.

### 4.2.1 Bit depth

Bit depth (the number of bits) refers to the number of CLs for each pixel (or channel in the case of color displays). Bit depth is described in powers of 2. Thus, a display with 256 commandable levels is said to be an 8-bit display. Bit depth ($BD$) is calculated as

$$BD = \frac{\log_{10} CL_{max}}{\log_{10} 2} \qquad (4.11)$$

and

$$CL_{max} = 2^{BD}, \qquad (4.12)$$

where $CL_{max}$ is the maximum command level (an integer). Bit depth describes the maximum number of unique command values that a pixel can assume. In the case of a color display, the number of bits per channel (levels per phosphor color) are added. An 8-bit per channel display is said to be a 24-bit display (3 channels × 8 bits/channel). A 24-bit display is theoretically capable of commanding 256 values[3] or 16,777,216 colors.

The number of uniquely commandable levels may not equal the number of unique luminance levels. For some monitors, the video driver may produce flat

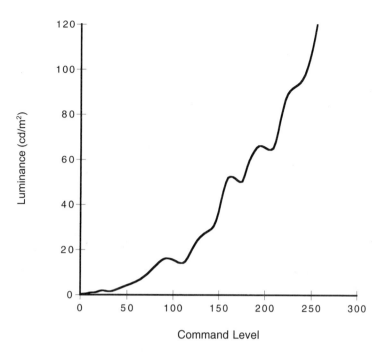

**Figure 4.12** Nonmonotonic CL/luminance function.

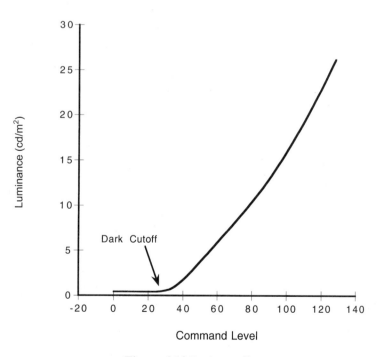

**Figure 4.13** Dark cutoff.

spots or reversals, as shown in Fig. 4.12. This is quite uncommon with current CRT monitors. Monitor luminance settings may also be such that the first several CL increases do not produce an increase in luminance (Fig. 4.13). Such a monitor is said to be in dark cutoff. Out-of-the box displays often exhibit this phenomenon. A similar phenomenon can occur at the top end of the luminance range as a monitor ages. More often, however, the monitor simply does not show monotonically increasing luminance with increasing CL. For this reason, bit depth is best measured by commanding all possible CLs and measuring the luminance output to determine the number of unique luminance steps.

### 4.2.2 Dynamic range

Although bit depth defines the number of potentially available gray tones or colors, it does not define the number that will actually be seen. The number of discernable luminance levels is a function of the luminance range, the minimum (Lmin) and maximum (Lmax) luminance levels, and the spacing of adjacent luminance levels.

The range of luminance values is called the dynamic range[11,12] (DR) and is defined as

$$DR = 10\log_{10}\frac{L_{max}}{L_{min}}. \qquad (4.13)$$

The DR is also sometimes expressed as the ratio of Lmax to Lmin, called the contrast ratio. Note that this range will vary depending on how it is measured. Internal monitor reflections, internal light scattering, and reflected light from the display room will reduce DR.

The HVS contrast discrimination ability varies with absolute luminance. As luminance increases, contrast sensitivity increases.[15] Given two displays with the same DR, the display with the higher value of Lmax (and thus Lmin) has potentially more discriminable contrast levels available. However, as will be shown subsequently, the distribution of luminance values over the DR also affects performance.

### 4.2.3 Gamma

Gamma is a measure often used to characterize the relationship between CL and luminance, typically plotted in log/log space.[5,16] Gamma is the slope of the function $\Delta\log_{10}L/\Delta\log_{10}CL$. Unfortunately, the measured function is often not linear so gamma depends on where the start and end points of the function are selected.[17] For that reason, gamma is not a definitive measure. Figure 4.14 shows a typical CRT gamma relationship.

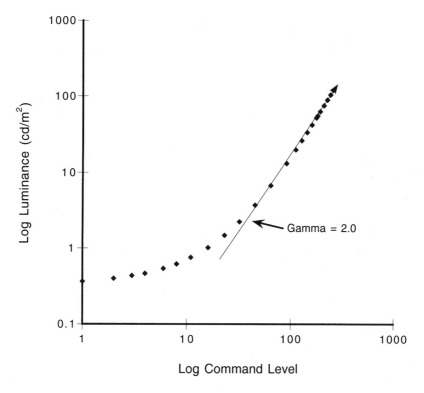

**Figure 4.14** A typical CRT gamma relationship.

### 4.2.4 Input/output function

The actual relationship between CL and luminance at each CL is a better characterization of monitor performance than gamma. The spacing of adjacent luminance levels is defined by the I/O function of the monitor.[18] The luminance difference between adjacent CLs generally increases as luminance (and CL) increases. The ability to see these differences is a function of both luminance and spatial characteristics. Models that were developed to predict performance of the HVS are described in Sec. 6.4. Look-up tables (LUTs) are used to modify the I/O function of the monitor to optimize contrast discrimination. The application of these LUTs (called perceptual linearization) ensures that each CL change produces an equally perceptible luminance difference.

### 4.2.5 Halation

A phenomenon called halation results in dark (low luminance) pixels that are surrounded by bright (high luminance) pixels showing higher luminance values than would be expected by their CL (Fig. 4.15). The luminance from the bright pixels "bleeds" into the dark area and thus increases the luminance of the surrounding pixels.

0% Halation                      8% Halation

**Figure 4.15** Halation.

### *4.2.6 Reflectance and transmittance*

External light reflects off the front of a display and adds to the light emitted by the display.[19, 20] Diffuse or Lambertian reflectance is direction independent in that the light strikes the face plate and diffuses in all directions. Specular reflections bounce the light back at an angle equal to the angle at which the light struck the face plate. Figure 4.16 illustrates the geometry of the two types of reflection. In actual fact, face plates are not perfect diffusers. A third aspect of reflection also exists called the haze component,[19] where a screen with a front surface treatment that diffuses light produces a specular reflection surrounded by haze.

It should be noted that light striking the display passes through the face plate and is reflected back through the face plate. Emitted light from the display passes through the face plate only once. Reflectance can be decreased by coating the face of the monitor with an antireflectance coating but at the expense of transmission through the face plate. Transmission is a measure of the proportion of light from the phosphor that passes through the face plate. By increasing transmission, Lmax can be increased without a significant increase in spot size. However, high-transmission face plates tend to have higher levels of halation and reflectance.

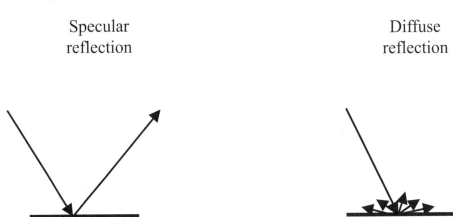

**Figure 4.16** Specular and diffuse reflection.

# PHYSICAL DISPLAY QUALITY MEASURES

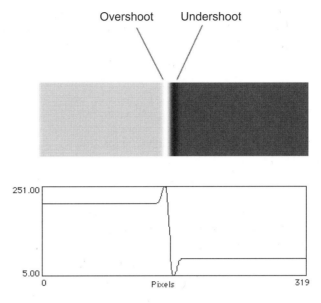

**Figure 4.17** Overshoot and undershoot.

## 4.2.7 Luminance stability

Luminance stability measures the degree to which maximum luminance varies as a function of the size of the commanded area. Full-screen luminance may be higher than that of small screen areas surrounded by a black background. The reverse may also occur, whereby luminance may increase as the size of the commanded area decreases. This is called luminance loading.[5] A measure called dynamic black-level stability measures the increase in luminance in a black area with a center white box. The white may "bleed" into the black, thus increasing luminance. Finally, video overshoot results in a change in pixel luminance values from their intended luminance values. As the electron beam sweeps across the display, the current level—and thus output luminance—may not change at a fast enough rate to keep up with changes in the image. Overshoot and undershoot occur when high frequencies are boosted to increase sharpness. This results in values "overshooting" or "undershooting" the intended value (Fig. 4.17).

## 4.2.8 Luminance and color uniformity

Luminance and color are not uniform across the face of a monitor. On a CRT, luminance typically falls off toward the edges of the display. Luminance nonuniformity is measured as a function of the maximum difference in luminance (%) at the maximum CL.[1] Differences of 20% are common. For an AMLCD, variations result from the pattern and diffusion of the backlighting source and from the angular dependence of the emitted light.

Color purity and uniformity can also vary over the face of the monitor. In addition, the color coordinates of the monitor white point may vary as a function of luminance level, called color tracking.[3] Gray areas on a monochrome image may appear to have a color cast at certain positions on the monitor or as intensity is varied.

### *4.2.9 Gamut*

Related to DR is the gamut of a color monitor. Although the eye is less sensitive to color differences than luminance differences, the range of possible colors (gamut) can affect the number of discernable colors. The gamut of a monitor is the range of color coordinate values that can be achieved by a monitor at Lmax. It is often expressed as a plot in some CIE coordinate space. Figure 4.18 shows two monitor gamuts in relationship to all possible colors. The gamut of a monitor is far smaller than the range of all possible colors. Various standards exist for gamuts (NTSC, SMPTEC, etc.), and they tend to trade increased brightness for decreased gamut.

The number of discernable color differences is also, in theory, affected by the color temperature of the monitor. Color temperature is defined by the color coordinate of the monitor white point. Increasing the temperature theoretically increases the ability to see yellow differences. It also tends to increase Lmax and thus the gamut.

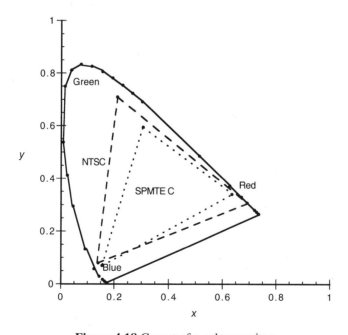

**Figure 4.18** Gamut of a color monitor.

## 4.2.10 Viewing angle

Cathode-ray tubes are relatively insensitive to the angle at which the display is viewed, but AMLCDs can be quite sensitive because of polarization effects. With LCD monitors, as one moves off center high luminance levels decrease but low luminance levels first increase and then decrease. This leads to gray-level reversal.[5] At some viewing angle, the contrast between two closely spaced gray levels decreases to 0 and then reverses. Similarly, chromaticity values change with viewing angle, and the change varies with brightness. This leads to actual changes in apparent color as a function of viewing angle.[21] A measure of the viewing angle is used to characterize this phenomenon where the viewing angle defines the angle over which some level of contrast or chromaticity constancy is maintained. Manufacturers tend to use the largest possible angle, e.g., the angle over which contrast is > 0. A more useful value is the angle at which the maximum contrast ratio (DR) is maintained above some specified value. A value of 50% is sometimes used.

## 4.3 Noise Measures

Noise, in the context of this display discussion, is defined as an unwanted variation in luminance strength or position resulting from performance of the video controller or monitor. The presence of noise reduces the ability to see small detail because the noise masks the contrast between the detail and its background.

### 4.3.1 Signal-to-noise ratio

Electronically, noise is measured as signal variation over time (Fig. 4.19). A variation in signal strength will produce a variation in output luminance. The ability to discern a signal difference is reduced by the amount of noise present. Noise is measured using the signal-to-noise ratio (SNR) defined as

$$\text{SNR} = \frac{S}{\sqrt{\sum_{i=1}^{n} \frac{(S_i - S_{avg})^2}{n}}}, \qquad (4.14)$$

where $S$ is the commanded or desired signal level, $S_i$ is the instantaneous signal level, and $S_{Avg}$ is the average signal level (Fig. 4.20). The SNR is typically measured in decibels:

$$\text{SNR}_{dB} = 20\log_{10}\text{SNR}. \qquad (4.15)$$

Noise is seldom as obvious on a good-quality image display monitor as it is on low-quality broadcast TV. Nonetheless, its presence interferes with the detection of

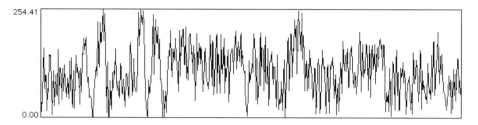

**Figure 4.19** Electronic noise.

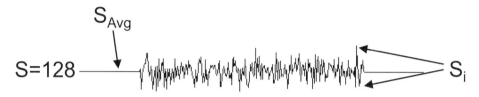

**Figure 4.20** SNR measurement.

**Figure 4.21** Example of noise.

small luminance differences and fine spatial detail. It can be seen as small variations in luminance when the display is commanded to the maximum command value. Figure 4.21 shows an example: A flat-field image of a screen has been enlarged and expanded over a larger DR. The actual range of CL values is 28, the standard deviation is 1.7, and the SNR is 38 dB. Noise occurs both spatially and over time. A given pixel does not exhibit the same luminance from frame to frame, and pixels with the same CL do not exhibit the same output luminance. However, the HVS also integrates over time and space, thus reducing the potential effects of noise.

Matrix displays exhibit another type of noise that results from pixel flaws. Due to manufacturing defects, a pixel may be always on or off, thus producing dark or light spots throughout the image that are continuously visible at the same position. Unless the number of spots is excessive, they are generally not a problem. However, if part of the pixel is on or off, a patterned noise may result. This can degrade the interpretability of imagery.

### 4.3.2 Noise power spectrum, noise-equivalent quanta, and detective quantum efficiency

Particularly in the medical display field, noise is measured and described in the Fourier domain. A fast Fourier transform (FFT) transforms the image from the spatial domain into the frequency domain. The FFT shows power (energy) as a function of spatial frequency. Figure 4.22 shows the FFT of a digital image of a flat-field CRT. Radial distance from the center of the image equates to spatial frequency and gray level to power. Thus, the image shows noise power as a function

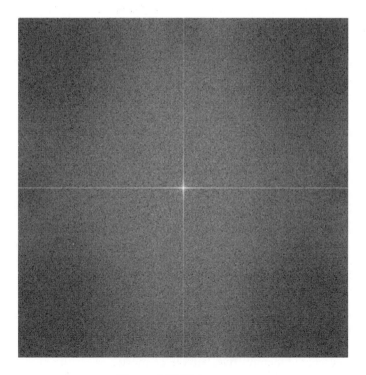

**Figure 4.22** Noise power.

of spatial frequency. The noise power is averaged in a circular fashion to produce a plot of noise power vs. spatial frequency or noise power spectrum (NPS). In some cases the data along the *x* and *y* axes are ignored.[22]

The ratio of the normalized MTF and NPS form a quantity called the noise-equivalent quanta (NEQ). In an ideal detector, every incident photon (energy unit) would be counted (detected). The NEQ is a measure of the effective number of photons counted by the device. If NEQ is divided by the number of incident quanta, we have a measure called the detective quantum efficiency (DQE). The DQE is a common measure in radiology, and with proper measurement it can be extended to show the contribution of the display.[22]

### *4.3.3 Jitter, swim, and drift*

On a CRT, the position of the raster can move over time. Movement is measured in terms of distance over time. Short-term (0.5 sec) movement is called jitter, intermediate-term (10 sec) movement is called swim, and long-term (30 sec) movement is called drift. Movement is measured in mils. One mil is 0.001 in.

### *4.3.4 Refresh rate and flicker*

If a display is not refreshed with sufficient frequency, it appears to flicker. The appearance of flicker varies with luminance and screen size, and among individuals. Figure 4.23 shows the rates above which flicker will not be perceived by 90% of the population.[23] On a 21-in. display, the corners of the display are at a 33-deg visual angle from a 16-in. viewing distance. The mean luminance of an aerial image is about 8 fL when viewed on a display with an Lmax of 35 fL. Radiographs and text tend to have higher mean values while MRI scans have lower values. A rate of 75 Hz is generally considered sufficient to avoid the appearance of flicker. With an AMLCD, the image is always lit so flicker is less likely to occur. The desired (or required) rate for AMLCDs and PDPs has not been defined.

### *4.3.5 Warm-up and aging*

Monitor performance may vary over time. As the monitor warms up, maximum performance is often not realized until 30 minutes or more after the monitor has been turned on. As a monitor ages, Lmax often decreases. This affects both DR and the monitor I/O function.

## 4.4 Artifacts and Distortions

Artifacts are unwanted features overlaid on an image as a result of the performance of the video driver, monitor, CPU, or connections among them. An image may be

PHYSICAL DISPLAY QUALITY MEASURES 75

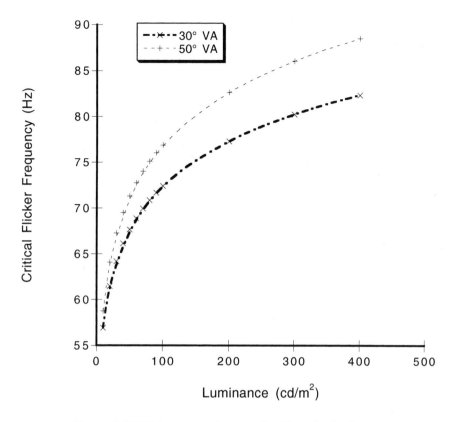

**Figure 4.23** Flicker perception as a function of refresh rate.

replicated and displaced in the scan direction as a ghost image; mismatched or overly long cabling is often the cause. Ghosting may also occur with an AMLCD. Ringing is an artifact that is related to overshoot and undershoot and results in signal variations after a transition (Fig. 4.24). Moiré is an aliasing artifact that results when the pixel pitch and spacing are mismatched. It occurs when the phosphor pitch is large relative to the pixel spacing. Moiré is illustrated in Fig. 4.25.

Blemishes may appear on AMLCDs as spots and streaks in a uniform gray field. The Japanese term "mura" is used for such defects. Mura is discussed in more detail in Chapter 9. Various other artifacts result from bandwidth compression, which are outside the scope of this book.

Geometric distortions occur with CRTs when the scan pattern is not regular and orthogonal, and significant distortions can be produced by errors in magnetic beam steering. Figure 3.20 illustrated various types of full-screen distortion. Monitors generally have controls to correct such distortions. The straightness (or waviness) of a display measures any nonlinearities over small areas of the display (Fig. 4.26). Measurements are made along three vertical and three horizontal lines at distances corresponding to 5% of the screen height or width. Straightness is defined as a percentage deviation relative to the full screen size. Thus, a deviation of 0.1 in. on a screen 10 in. high would be reported as a 1% deviation. Linearity

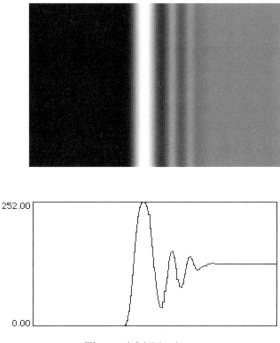

**Figure 4.24** Ringing.

**Figure 4.25** Moiré.

measures the relationship between the commanded and actual location of pixels on a display. It is defined, in the same manner as straightness, as a percentage deviation relative to the full screen size. It is measured using a vertical and horizontal grill with line spacings equal to 5% of the screen dimension.[5]

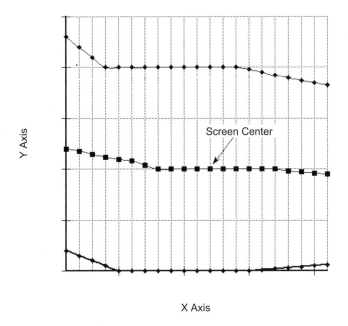

**Figure 4.26** Straightness measurement.

On some CRTs, the size of the image can vary as a function of image luminance. As screen luminance increases, the size of the raster can decrease. A measure of this increase relative to some lower defined luminance level is called raster stability. If the monitor power supply is well regulated, there should be a negligible increase in raster size.

## 4.5 Categorization by Measurement Domain

In the previous four sections, physical quality measures were defined in terms of the quality domain; e.g., resolution and contrast. The same measurements can also be categorized in terms of their measurement domain—luminance, spectral, spatial, and temporal (time). These categories imply the types of measurement equipment required. Luminance measures display performance related to brightness, and in most cases requires a photometer for measurement. In other cases—e.g., moiré—the presence or absence of a quality value or defect is defined visually or by calculation. Spectral measures relate to color and require a colorimeter or spectroradiometer for measurement. Spatial measures include both resolution and geometric distortion that require measuring devices such as rulers, photometers, and CCD arrays. Temporal measures define luminance or spatial variations as a function of time. This categorization of measurement domains is summarized in Table 4.1.

**Table 4.1** Measurement categorization.

| Domain | Measure |
|---|---|
| Luminance response | DR <br> Lmax <br> Lmin <br> Bit depth <br> Gamma <br> I/O function <br> Halation <br> Reflectance and transmittance <br> Cm <br> SNR/NPS/NEQ <br> Ringing <br> Streaking and ghosting <br> Mura <br> Luminance stability <br> Luminance uniformity <br> Moire' <br> Acceptable viewing angle |
| Spectral response | Color temperature <br> Color uniformity <br> Color tracking <br> Gamut <br> Acceptable viewing angle |
| Spatial | Size of viewable area <br> Addressability <br> Pixel density <br> Pixel pitch <br> Pixel size/line width <br> RAR <br> RER <br> Edge response <br> MTF <br> DQE <br> Pixel subtense <br> Angular resolution <br> Effective resolution <br> Straightness <br> Linearity <br> Raster stability |
| Temporal | Flicker <br> Jitter <br> Swim <br> Drift <br> SNR <br> Bandwidth <br> Warm-up time |

## 4.6 Summary

Physical measures of display quality define the spatial and spectral accuracy of the displayed image relative to the imaged scene or object. Measures are categorized by resolution, the ability to see spatial detail; and by contrast, the ability to detect tone or color differences. The two types of measures overlap in that fine detail cannot be seen if contrast is not sufficient, and contrast discrimination varies as a function of resolution. Measures are also categorized by noise (signal variation over time or position), artifacts, and geometric distortions. Noise also reduces the ability to see small luminance differences.

Physical quality measures can be described in terms of their measurement domain. Measurement domains include luminance, spectral, temporal, and spatial. The measurement domains in turn define the measurement instruments required.

## References

[1] National Information Display Laboratory, *Display Monitor Measurement Methods under discussion by EIA Committee JT-20, Part 1, Monochrome CRT Monitor Performance,* Version 2.0, Princeton, NJ (1995).

[2] S. Wearden, "Landscape vs. portrait formats," *Future of Print Media,* on-line journal, *Institute for Cyberinformation,* Kent State University, http://www.futureprint.kent.edu/about.htm (1998).

[3] P. A. Keller, *Electronic Display Measurement,* John Wiley & Sons, Inc., New York (1997).

[4] National Information Display Laboratory, *Display Monitor Measurement Methods under Discussion by EIA, Part 2, Color CRT Monitor Performance,* Version 2.0, Princeton, NJ (1995).

[5] Video Electronic Standards Association, *Flat Panel Display Measurements Standard,* Version 2.0, VESA, Milpitas, CA (2001).

[6] G. M. Murch and R. J. Beaton, *Relating Display Resolution and Addressability,* TEB27, Electronic Industries Association, Arlington, VA (1988).

[7] J. C. Leachtenauer and R. G. Driggers, *Surveillance and Reconnaissance Imaging Systems: Modeling and Performance Prediction,* Artech House, Boston, MA (2001).

[8] R. Donofrio, "Do direct-view CRTs make sense for HDTV?" *Information Display,* June 1999, pp. 22–25.

[9] J. Leachtenauer, W. Malila, J. Irvine, L. Colburn, and N. Salvaggio, "General Image-Quality Equation: GIQE," *Applied Optics,* Vol. 36(32), pp. 8322–8328 (1997).

[10] Eastman Kodak Co., *Quality Assessment Report for Imaging Displays,* Vol. 1(2), 1998.

[11] J. Leachtenauer, *SC Measurement Procedures,* NIMA, Reston, VA (1998).

[12] National Information Display Laboratory, *Request for Evaluation Monitors for the National Imagery and Mapping Agency (NIMA) Integrated Exploitation Capability (IEC),* NIDL at Sarnoff Corporation, Princeton, NJ (1999).

[13] H. Roehrig, "The monochrome cathode ray tube display and its performance," in Y. Kim and S. Horii (eds.), *Handbook of Medical Imaging, Vol. 3, Displays and PACs*, SPIE Press, Bellingham, WA, pp. 157–220 (2000).

[14] R. J. Farrell and J. M. Booth, *Design Handbook for Imagery Interpretation Equipment*, Boeing Aerospace Co., Seattle, WA (1984).

[15] P. J. G. Barten, *Contrast Sensitivity of the Human Eye and Its Effect on Image Quality*, SPIE Press, Bellingham, WA (1999).

[16] National Information Display Laboratory, *Test Procedures for Evaluation of CRT Display Monitors,* Version 2.0, Princeton, NJ (1991).

[17] E. Muka, H. Blume, and S. Daly, "Display of medical images on CRT softcopy displays: a tutorial," *Proc. SPIE*, Vol. 2431, pp. 341–358 (1995).

[18] M. Weibrecht, G. Spekowius, P. Quadfleig, and H. Blume, "Image quality assessment of monochrome monitors for medical softcopy display," *Proc. SPIE*, Vol. 3031, pp. 232–244 (1997).

[19] E. F. Kelley, G. R. Jones, and T. A. Germer, "The three components of reflection," *Information Display*, Vol. 10, pp. 24–29 (1998).

[20] M. H. Brill, "Seeing through screen reflection," *Information Display*, Vol. 1, pp. 28–31 (1999).

[21] M. H. Brill, "Color reversal at a glance," *Information Display*, Vol. 6, pp. 36–37 (2000).

[22] J. T. Dobbins, "Image quality metrics for digital systems," in J. Beutel, H. L. Kundel, and R. L. Van Metter (eds.), *Handbook of Medical Imaging, Vol. 1, Physics and Psychophysics,* SPIE Press, Bellingham, WA, pp. 3–78 (2001).

[23] J. E. Farrell, B. L. Henson, and C. R. Haynie, "Predicting flicker thresholds for video display terminals," *Proceedings of the Society for Information Display*, Vol. 28-4, pp. 449–453 (1987).

# Chapter 5
# Perceptual Quality and Utility Measures

The literature on display quality and utility deals with both physical and perceptual measures. The previous chapter defined and described physical measures; this chapter describes common perceptual measures. Later chapters will show the results of studies using these measures in order to demonstrate the effects of display parameter differences.

Perceptual measures are those that involve the HVS and thought processes (known as the cognitive system). These measures are used to assess the quality or utility of a display. Whereas physical quality measures are made at the output of the display, perceptual quality measures include the human visual and cognitive systems of the image chain (Fig. 1.1). It is sometimes the case that the HVS is the limiting factor in the display chain. For example, under normal viewing conditions, the observer may not be able to resolve all of the low-contrast detail presented on the face of the display.

The purpose of a display is to present information for analysis and decision making. The quality of the information presented must be sufficient to enable accurate analysis and decision making in a timely manner. The value of the information presented defines utility. At some point on a quality continuum, there may be no improvement in utility. The relationship between quality and utility is non-monotonic as shown in Fig. 5.1.[1] Measurement of display utility involves both the visual and cognitive systems.

Perceptual measures of quality and utility range from subjective ratings of display quality (e.g., poor, fair, good) to measures of task performance using the display (e.g., detection of lesions on lung x rays). At an intermediate level, observers may make estimates of their ability to perform defined tasks on a display. The National Imagery Interpretability Rating Scale (or NIIRS, commonly pronounced "nears") is an example of a task performance estimate scale.[2] Although physical and perceptual measures are related, they do not always relate in a linear fashion. The HVS has finite limits—we cannot see infinitely small detail. Thus, improving resolution beyond our ability to see the image detail can reduce the usefulness of a display.

Similarly, improving some measure of physical quality does not necessarily improve perceptual quality. Figure 5.2 provides an example. As gamma increases, perceptual contrast increases, but judgements of image quality do not vary accordingly.[3]

This chapter will describe both subjective and objective measures of quality and utility. Subjective quality measures require observers to provide judgements

**Figure 5.1** Utility vs. quality.

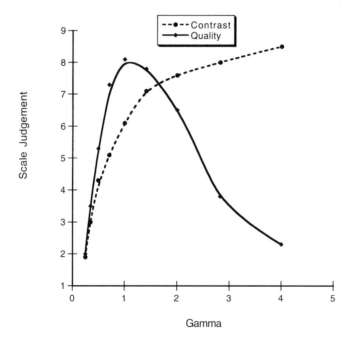

**Figure 5.2** Perceptual quality vs. gamma. Data from Ref. [3] used with permission.

on the relative or absolute quality of a displayed image. Subjective performance or utility estimates require observers to estimate their ability to perform a specific task on the image displayed. Objective quality measures require observers to respond to

## 5.1 Subjective Quality Ratings

Subjective quality ratings generally entail some type of category scale rating. They may be binary (better/worse) or they may involve some type of category scale. Categories might, for example, range from good to poor on a five-point scale. Subjective quality ratings are typically used to reach conclusions regarding the effects of alternative image processing algorithms or levels. For example, different levels of bandwidth compression might be applied to images and observers asked to make subjective judgments as to the quality of the processed images. Results would then be analyzed to determine at what level the compression was or was not perceptually acceptable. Studies of television quality often employ what is called the double stimulus continuous quality scale (DSCQS) and the double stimulus impairment scale (DSIS).[4] The DSCQS and DSIS are shown in Figs. 5.3 and 5.4. In both cases, the video sequence to be rated is viewed along with a standard image. The image sequences are viewed alternately and the rating made after one or more presentations of the image sequences. The DSCQS and DSIS follow rather rigorous methods of presentation and data analysis.

Similar five- to ten-point scales can be used for a variety of purposes and may have anywhere from two to five verbal anchors (verbal descriptions associated with numerical points on the scale). Figure 5.5 illustrates a scale used to rate the relative quality of a color display used to view color imagery. Such scales can

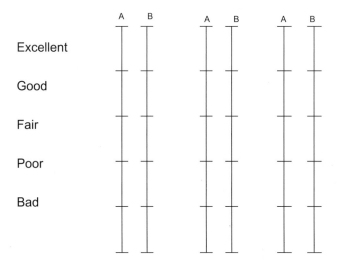

**Figure 5.3** DSCQS rating form. Reprinted with permission from *Surveillance and Reconnaissance Imaging Systems: Modeling and Performance Prediction*, by Jon C. Leachtenauer and Ronald G. Driggers, Artech House Inc., Norwood, MA, USA (www.artechhouse.com).

also be used to rate specific aspects of display quality such as sharpness, contrast, and noise.

## 5.2 Subjective Performance (Utility) Estimates

One difficulty with subjective quality ratings is that results are relative and cannot directly be tied to measures of performance—some type of calibration is required. In the case of the double stimulus scales, the rating is tied to the quality of the comparison image or sequence. A DSCQS rating of "good" is tied to the quality of the original used as the basis for comparison. For binary scales, image A may be better than image B, but the practical significance of the difference is unknown.

A way of at least partly overcoming this problem is to ask for estimates of task performance. A common method is to elicit confidence ratings: "I am x% confident I could perform the task on a display of this quality." Another method is to ask for a binary decision along with a confidence rating: "I believe (with 70% confidence) that there is a (tank/target/lesion) present on this display."

The assumption is that the confidence ratings predict performance, and this, in fact, has been shown to be true for positive (correct response) performance.[5, 6] Observers cannot accurately predict negative performance.[5]

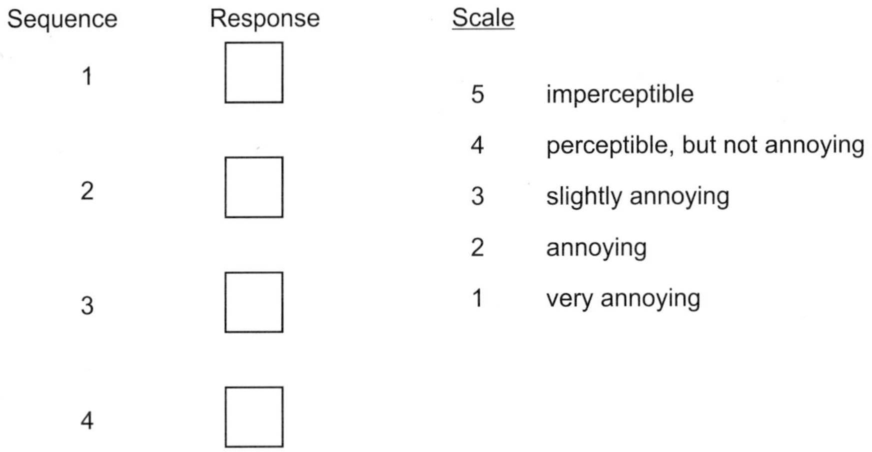

Sequence
Response
Scale
1
2
3
4
5 imperceptible
4 perceptible, but not annoying
3 slightly annoying
2 annoying
1 very annoying

**Figure 5.4** DSIS rating form. Reprinted with permission from *Surveillance and Reconnaissance Imaging Systems: Modeling and Performance Prediction*, by Jon C. Leachtenauer and Ronald G. Driggers, Artech House Inc., Norwood, MA, USA (www.artech-house.com).

-10 Significantly Worse
-5 Moderately Worse
0 Equal
5 Moderately Better
10 Significanlty Better

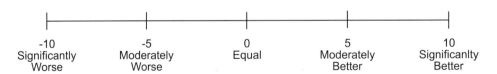

**Figure 5.5** Subjective color image relative quality rating scale.

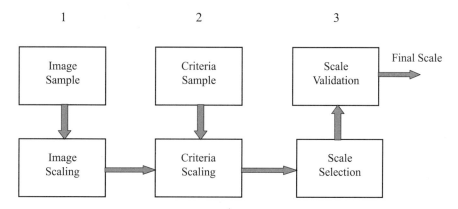

**Figure 5.6** NIIRS development process.

## 5.3 National Imagery Interpretability Rating Scale

The NIIRS is termed a subjective performance estimate scale. A rating represents the rater's judgment of the most difficult task that can be performed on an image of the displayed quality.[2] The NIIRS was developed as a replacement for subjective relative quality scales (which use ratings of poor to excellent). A NIIRS is shown in Table 5.1 for visible spectrum imagery. Similar scales exist for infrared (IR) radar and multispectral imagery. The NIIRS is used extensively in studies of surveillance and reconnaissance systems, and in the tasking and performance assessment of such systems.

The NIIRS is developed in a three-step process (Fig. 5.6).[7] First, a large sample of imagery is scaled on a relative basis (0–100 scale) in terms of what is called image interpretability. Interpretability is defined as the ability to extract intelligence information from an image. Higher levels of image quality provide more information (or information of a more detailed nature). The image scaling is performed by imagery analysts experienced with the type of imagery under consideration (e.g., radar). The image sample is typically grouped by content category such as ships, aircraft, and vehicles. Scaling is performed separately for each grouping. The results are used to form a subjective quality scale (SQS) for each imagery grouping.

Marker images are selected at the quartile (0, 25, 50, 75, 100) or quintile level (0, 20, 40, 60, 80, 100) to serve as the basis for the next step of the process, criteria scaling. Many interpretability tasks are defined for each content category grouping. They range in difficulty from very easy (in terms of the required level of image/display quality) to very difficult. These tasks are rated on the SQS defined by the previous image scaling task. Again, qualified imagery analysts determine the ratings. For example, if the analyst believes a task can be performed on the 80 marker image but not the 60, the task would receive a rating somewhere between 61 and 80. At the completion of this step, each task has an assigned level of required image interpretability.

### Table 5.1 Visible NIIRS.

**Visible National Image Interpretability Ratings Scale-March 1994**

#### Rating Level 0
Interpretability of the image is precluded by obscuration, degradation, or very poor resolution.

#### Rating Level 1
Detect a medium-sized port facility and/or distinguish between taxiways and runways at a large airfield.

#### Rating Level 2
Detect large hangars at airfields.
Detect large static radars (e.g., AN/FPS-85,COBRA DANE, PECHORA, HENHOUSE)
Detect military training areas.
Identify an SA-5 site based on road pattern and overall site configuration.
Detect large buildings at a naval facility (e.g., warehouses, construction hall).
Detect large buildings (e.g., hospitals, factories)

#### Rating Level 3
Identify the wing configuration (e.g., straight, swept, delta) of all large aircraft(e.g.,707,CONCORD,BEAR,BLACKJACK).
Identify radar and guidance areas at a SAM site by the configuration, mounds, and presence of concrete aprons.
Detect a helipad by the configuration and markings
Detect the presence/absence of support vehicles at a mobile missile base.
Identify a large surface ship in port by type (e.g., cruiser, auxillary ship, noncombatant/merchant)
Detect trains or strings of standard rolling stock on railroad tracks (not individual cars)

#### Rating Level 4
Identify all large fighters by type (e.g., FENCER,FOXBAT,F-15,F-14)
Detect the presence of large individual radar antennas (e.g., TALL KING)
Identify, by general type, tracked vehicles, field artillery, large river crossing equipment, wheeled vehicles when in groups.
Detect an open missile silo door.
Determine the shape of the bow (pointed or blunt/rounded) on a medium-sized submarine (e.g., ROMEO,HAN, Type 209, CHARLIE II, ECHO II, VICTOR II/III)
Identify individual tracks, rail pairs, control towers, switching points in rail yards.

#### Rating Level 5
Distinguish between a MIDAS and a CANDID by the presence of refueling equipment (e.g., pedestal and wing pod)
Identify radar as vehicle-mounted or trailer-mounted.
Identify, by type, deployed tactical SSM systems (e.g. FROG, SS-21, SCUD).

#### Rating Level 5 (cont.)
Distinguish between SS-25 mobile missile TEL and Missile Support Vans (MSVs) in a known support base, when not covered by camouflage.
Identify TOP STEER or TOP SAIL air surveillance radar on KIROV-, SOVREMENNY-KIEV-SLAVA-MOSKVA-KARA-, or KRESTA-II-class vessels.
Identify individual rail cars by type (e.g., gondola, flat, box) and/or locomotives by type (e.g., steam, diesel).

#### Rating Level 6
Distinguish between models of small/medium helicopters (e.g., HELIX A from HELIX B from HELIX C, HIND D from HIND E, HAZE A from HAZE B from HAZE C).
Identify the shape of antennas on EW/GCI/ACQ radars as parabolic, parabolic with clipped corners, or rectangular.
Identify the spare tire on a medium sized-truck.
Distinguish between SA-6, SA-11, and SA-17 missile airframes.
Identify individual launcher covers (8) of vertically launched SA-N-6 on SLAV-class vessels.
Identify automobiles as sedans or station wagons.

#### Rating Level 7
Identify fitments and fairings on a fighter-sized aircraft (e.g., FULCRUM, FOXHOUND).
Identify ports, ladders, vents on electronic vans.
Detect the mount for antitank guided missiles (e.g., SAGGER on BMP-1).
Detect details of the silo door hinging mechanism on Type III-F, III-G, II-H launch silos and type III-X launch control silos.
Identify the individual tubes of the RBU on KIROV-, KARA-, KRIVAK-class vessels.
Identify individual rail ties.

#### Rating Level 8
Identify the rivet lines on bomber aircraft.
Detect horn-shaped and W-shaped antennas mounted atop BACK TRAP and BACKNET radars.
Identify a hand-held SAM (e.g., SA-7/14,REDEYE, STINGER).
Identify joints and welds on a TEL or TELAR.
Detect winch cables on deck-mounted cranes.
Identify windshield-wipers on a vehicle.

#### Rating Level 9
Differentiate cross-slot from single slot heads on aircraft skin panel fasteners.
Identify small, light-tones ceramic insulators that connect wires of an antenna canopy.
Identify vehicle registration numbers (VRN) on trucks.
Identify screws and bolts on missile components.
Identify braid of ropes (three to five inches in diameter).
Detect individual spikes in railroad ties.

The final step in the development process is to select the criteria that define the NIIRS and then to validate the scale. The criteria are selected at equal intervals across the SQS such that nine task criteria are defined for each content category. Normally, this would entail selecting criteria rated at multiples of 11. Other factors, such as low rating variability, enter into the final selection. The validation process entails imagery analyst (IA) ratings of a sample of imagery to ensure that the scale has the desired properties such as linearity (with respect to SQS) and separability (images that differ by a full NIIRS level should be statistically separable).

Although the NIIRS does not directly measure image quality, NIIRS ratings vary as a function of physical image quality measures. For example, under some circumstances, for each doubling of image resolution, the NIIRS changes by one level.[8] The relationship between NIIRS and resolution is thus logarithmic (base 2). This factor is key to interpreting NIIRS differences. Thus, a loss of 0.2 NIIRS represents a resolution loss of 15% (all other things being equal). To make up this loss for a given system requires that the collector operate at a 15% lower altitude. This in turn requires more images to cover the same ground area. The increase is 38%, meaning that 138 images must be acquired for every 100 images at the higher altitude. This, of course, translates to more time. Other factors such as noise and edge sharpness, however, also affect interpretability and thus NIIRS ratings.

The variability in NIIRS ratings is described in terms of the standard deviation of ratings made by several raters on a single image. A standard deviation of 0.5 to 0.6 is typical for NIIRS criteria ratings. NIIRS difference rating variability (ratings made by expressing a NIIRS difference between two images) is lower, typically on the order of 0.2 to 0.3.

## 5.4 Objective Perceptual Quality Measures

Objective perceptual quality measures use observer ratings of specially designed targets to measure image and display quality. For many years, various types of resolution targets were used to assess image quality. Figure 5.7 illustrates several resolution targets, ranging from the Snellen letter to the tri-bar. Perhaps the most commonly used target in the image community until the 1970s was the USAF tri-bar target (Fig. 5.8).[9] The observer determined the smallest vertical and horizontal pattern in which the three bars could be separately distinguished (resolved). The ability to resolve closely spaced objects is a function not only of size and spacing, but of contrast as well. The tri-bar target thus defined a measure of resolution only for the contrast level at which it was presented. It was often the case that the contrast of the target was not specified, which meant that results could be related only to the undefined target. In some cases, the contrast of the original target was specified, but often at a level that was unrealistic in terms of normal scene content. Targets with a ratio of 1000:1 were used, but a ratio of 1.6:1 was considered more realistic in terms of visible imagery scene content.[10] Tri-bar targets also suffered from other problems such as spurious resolution related to sampling artifacts. Despite these problems, tri-bar resolution had the advantage of simplicity of measurement and

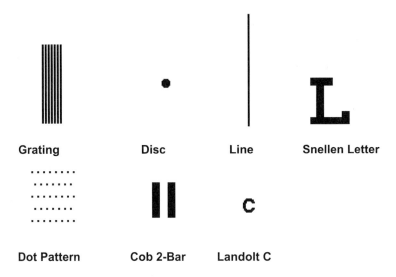

**Figure 5.7** Resolution targets.

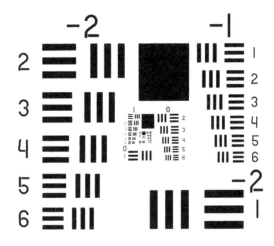

**Figure 5.8** USAF tri-bar target.

communication.[11] More "pure" measures such as the MTF were more difficult to measure and could not be reduced to a single meaningful value.

In a study that compared a variety of resolution targets, a checkerboard target was found to be the most reliable of the targets tested. Further, a checkerboard target can be constructed in a smaller space than a tri-bar target. With these facts as a basis, Dr. S.J. Briggs of the Boeing Company developed what is called the Briggs target.[12]

## 5.4.1 Briggs target

The Briggs target is a series of checkerboard targets that differ in size, contrast, and absolute luminance values (when displayed). The target thus measures the ability to resolve fine detail as a function of contrast. Figure 5.9 illustrates a Briggs target set, which is made up of two or more Briggs targets. The individual Briggs target, shown in Fig. 5.10, consists of 17 checkerboards. The checkerboards vary in the

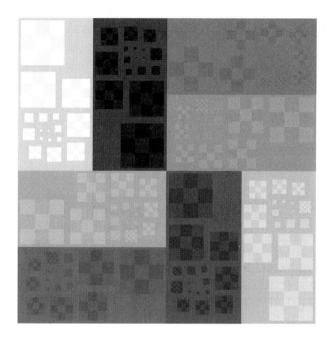

**Figure 5.9** Briggs target set.

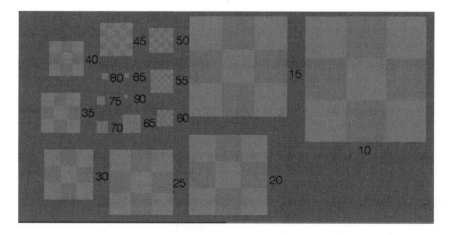

**Figure 5.10** Briggs target.

number of pixels per square (called "checkers" by Briggs) and the number of squares per board. The light and dark squares differ by a specified number of CLs; the background CL is a function of the values of the squares. The background is defined as

$$B_i = \frac{[0.7(L_i + D_i) + 0.3(N - 1)]}{2}, \quad (5.1)$$

where $L_i$ and $D_i$ are the CL values of the light and dark squares, respectively, 0.3 is an empirically defined value, and $N$ is the number of gray levels.[13]

The Briggs target set typically contains eight targets arranged as shown in Fig. 5.9. Target sets are defined in terms of the CL difference between the dark and light squares. The common values are 1, 3, 7, and 15, which are labeled the C-1, C-3, C-7, and C-15 target sets. The targets within a set are equally spaced across the CL range (see Table 5.2). The targets are arranged in terms of CL and luminance as shown in Table 5.2 and Fig. 5.9. In showing the results of a Briggs rating assessment, it is usually more meaningful to display results in terms of increasing luminance/CL order rather than the order shown in Table 5.2.

A 15-target set has also been produced with the additional seven targets falling halfway between the original eight. Since the size of the squares in the Briggs target differ, Cm will also differ. The Briggs target can therefore be considered a measure of Cm performance.

Table 5.2 Briggs target CL values for C-7 target.

| Target | Brightness rank | Light CL | Dark CL | Background CL |
|---|---|---|---|---|
| T-1 | 8 | 255 | 248 | 214 |
| T-2 | 1 | 7 | 0 | 41 |
| T-3 | 4 | 113 | 106 | 115 |
| T-4 | 5 | 149 | 142 | 140 |
| T-5 | 7 | 220 | 213 | 189 |
| T-6 | 2 | 42 | 35 | 65 |
| T-7 | 3 | 78 | 71 | 90 |
| T-8 | 6 | 184 | 177 | 165 |

The designation and size of the checkerboards vary as shown in Table 5.3. The number of squares in the Briggs targets was determined empirically. Since a digital display must be defined in pixel units, the perceptual change as the number of pixels per square decreases becomes larger and larger (e.g., the perceptual difference between one and two pixels is much larger than that between three and four). To overcome this problem, the number of squares per board was also varied. The final dimensions were selected to provide roughly equal steps in perceptual difficulty. A function was fit to a set of data of the form

$$Y = k_1 \log_{10} \frac{1}{p} + k_2 \log_{10} \frac{1}{B-1} + k_3, \tag{5.2}$$

where $p$ is the number of pixels in a square and $B$ is the number of squares in a board.

It should be noted that the Briggs target development data were obtained using a 1977-era color shadow mask monitor with an addressability of approximately 50 ppi. Luminance measurements reported suggested that the I/O function was not monotonic.[12]

**Table 5.3** Briggs target spatial characteristics.

| Board # | Pixels/Square | Squares/Row |
|---|---|---|
| B-10 | 25 | 3 |
| B-15 | 20 | 3 |
| B-20 | 16 | 3 |
| B-25 | 13 | 3 |
| B-30 | 10 | 3 |
| B-35 | 8 | 3 |
| B-40 | 7 | 3 |
| B-45 | 4 | 5 |
| B-50 | 3 | 5 |
| B-55 | 2 | 7 |
| B-60 | 2 | 5 |
| B-65 | 1 | 11 |
| B-70 | 1 | 7 |
| B-75 | 1 | 5 |
| B-80 | 1 | 4 |
| B-85 | 1 | 3 |
| B-90 | 1 | 2 |

The Briggs target is currently scored by recording the number of the smallest board in which the individual squares can be discerned. It is generally more convenient to number the boards 1 through 17 instead of the B-10 through B-90 designations. The relative sharpness of the squares in the smallest resolvable board is scored using the scale shown in Fig. 5.11. A square that is sharp and well defined receives a score of 1 and a "blob" a score of 5. The overall Briggs score is defined by

$$\text{Score} = [(\text{Target \#} \times 5) + 6] - \text{Sharpness Value}. \tag{5.3}$$

In Fig. 5.12, the eleventh largest board is evident and the smaller boards are not. If the squares in the board had a rating of 3, the Briggs score for this target would be

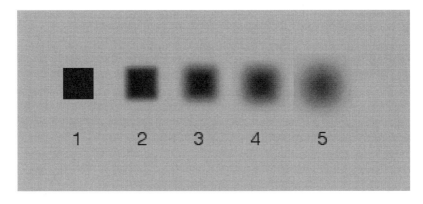

**Figure 5.11** Target rating scale.

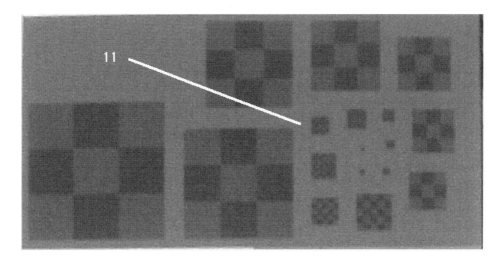

**Figure 5.12** Briggs target example.

58 (11 × 5 + 6 = 61, 61 – 3 = 58). Scores can be plotted for each individual target or can be averaged across the targets in a target set.

The Briggs target can be used in a variety of ways. If the goal is to define the physical quality of a display (independent of observer limitations), the observer should be allowed to use optical magnification in making a rating. If the goal is to define observer perception, no magnification should be used. Observers are, however, encouraged to move close to and away from the display in order to maximize their readings (the larger boards are sometimes more evident when viewing distance is increased). The ability to detect small contrast differences is greatest at a particular spatial frequency (~ 2 cy/deg). Varying the distance to the display allows the observer to bring a particular Briggs pattern into the optimum spatial frequency range (within the limits of vision).

The Briggs target is most reliable when readings average around 50, i.e., when the largest boards are discernible and the smallest are not. No contrast reversals should be evident. This is seldom a problem with current displays, with the exception of LCDs viewed from a nonorthogonal direction.

Often the C-1 target is not discernible but all of the boards in the C-15 target are discernible. This means the C-3 and C-7 targets are generally the most reliable. Rater variability is on the order of 6 to 8 for normal situations, with normal defined as cases where all eight targets are evident with about the same level of clarity. Variability decreases when observers are allowed to use magnification (i.e., the display and not the observer-display combination is measured). Variability increases when there is a large age disparity among the observers.

Briggs scores can be normalized to an adjusted score by the contrast level using the function

$$\text{Normalized Score} = \text{Score} \times \log\left(\frac{N-C}{C}\right), \quad (5.4)$$

where $N$ is the number of CLs (256 for an 8-bit system) and $C$ is the contrast level of the target (1, 3, 7, or 15). Note that these scoring procedures differ slightly from those in the original Briggs manual.[13]

A color Briggs target has also been developed.[14] Figure 5.13 shows one color target in monochrome (the color targets are on the enclosed CD). The general principle behind the color target is the same as the monochrome Briggs. The color Briggs is defined in CIE $L*u*v*$ space with three targets centered on the blue-green, yellow-red, and yellow-green axes of the monitor gamut (Fig. 5.14). The design goal was to obtain scores of ~ 50 when the targets were viewed at 2x magnification. Four roughly equally spaced L* values were defined and ΔC (chroma) values defined empirically using data from three image scientists. Because of the desire to achieve scores of 50, the ΔC values varied as a function of the color and

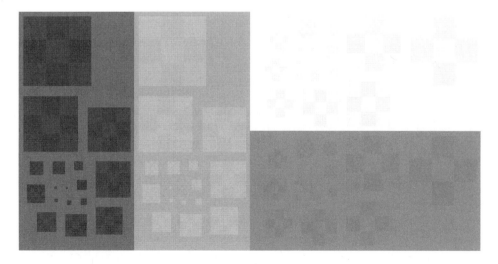

**Figure 5.13** Color Briggs target (color targets shown on enclosed CD).

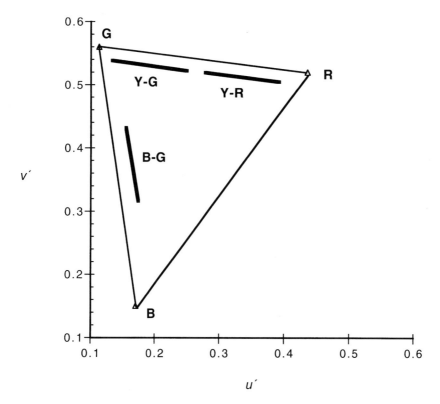

**Figure 5.14** Location of color Briggs target in CIE space.

L* level. Table 5.4 shows the actual ΔC values. Subsequent testing showed that the ΔC values for the yellow-green target were too small (i.e., about 8–14 points lower than those of the other two targets) as shown in Fig. 5.15. Variability of the color target was 6.7 units in one study.[14]

**Table 5.4** Color Briggs delta-C values.

| L* Level | Y-G Target | Y-R Target | B-G Target |
|---|---|---|---|
| 1 | 17 | 18 | 38 |
| 2 | 13 | 13 | 21 |
| 3 | 7 | 7 | 40 |
| 4 | 18 | 18 | 18 |

A comparison of two objective quality measures is provided by a study in which Briggs target ratings and a four-alternative forced choice (4-AFC) technique were compared. The 4-AFC metric required observers to determine the location (four possible positions) of a low-contrast square embedded in a larger square. Both the Briggs targets and the 4-AFC targets were processed to different MTF levels. The Briggs target was also processed at different noise levels. Results indicated the 4-

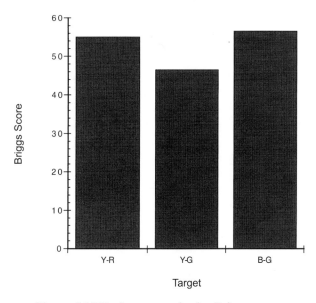

**Figure 5.15** Performance of color Briggs target.

AFC metric slightly outperformed the Briggs measure. The Briggs variability was larger than has been observed in other studies. Neither of the metrics was considered successful in detecting MTF differences.

### 5.4.2 Briggs vs. NIIRS

Since the Briggs target measures the ability to resolve detail and the NIIRS is strongly influenced by resolution, one would expect a correlation between Briggs ratings and NIIRS ratings. Such a correlation does exist, but it is nonlinear. Several studies have been performed in which both NIIRS and Briggs ratings were acquired. The NIIRS ratings were defined and expressed as NIIRS difference values (delta-NIIRS). Results from three of these studies are shown in Fig. 5.16.[14, 16, 17] It should be noted that the Briggs scores for studies 1 and 2 were the average rating on the C-7 target made by 8–10 imagery analysts with unaided viewing. Scores for study 3 were obtained using the color Briggs target. Scores reflect limitations of both the monitor and the HVS. The studies used different monitors with different magnification and pixel density values (50, 72, and 100 ppi). Bilinear interpolation was used for magnification. This had the effect of avoiding blocking—which occurs when the same pixel is replicated four times (for 2x magnification) and the image appears to be made up of "blocks" as opposed to single pixels—but also decreased Briggs scores relative to the use of pixel replication. In one of the studies cited,[17] the Briggs score increased by one point as a function of magnification; it would have increased by about 15 points with pixel replication. Finally, to normalize pixel density differences across monitors, pixel density differences were adjusted based on pixel angular subtense.

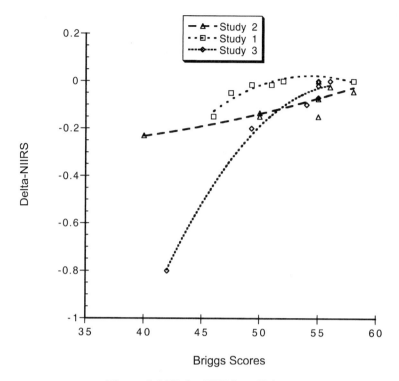

**Figure 5.16** Delta-NIIRS vs. Briggs.

The study results indicated a strong relationship between the Briggs and NIIRS ratings. Studies 1[14] and 3[17] showed a nonlinear relationship while study 2[16] showed a linear relationship. All three studies, however, indicated a loss in NIIRS when the C-7 or color Briggs score fell below 50 to 55. Scores of 55 and 60 indicate the ability to resolve two pixel squares. When a subset of the data from study 2 was replotted (restricted to monitors with 100 ppi original pixel density), the data suggested that performance improved up to a score of 60–65 (one pixel). The data are shown in Fig. 5.17.

Data from another study indicate that at some point the eye becomes the limiting factor.[18] This study showed that Briggs scores increased with decreasing pixel density, and the increase was linear with the log of pixel density (provided that pixel replication was used). Figure 5.18 shows the data. NIIRS scores, however, peaked at some intermediate level of pixel density (Fig. 5.19).

### 5.5 Objective Performance (Utility) Measurement

Arguably the most rigorous approach to perceptual quantification of quality is the use of objective performance measures. We say "arguably" because some of these measures are not easy to use and do not necessarily provide a single answer. Objective performance measures require the observer to state that a "target" is or is not

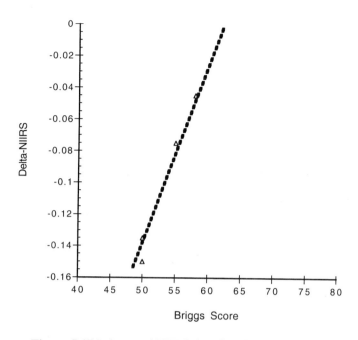

**Figure 5.17** Briggs and NIIRS data for 100-ppi monitors.

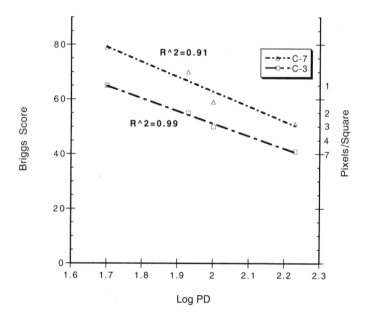

**Figure 5.18** Effect of pixel density on Briggs C-7 scores.

present (detect a target) or to identify a detected target to some defined level of specificity. Results are typically expressed as probabilities of task performance. Specificity of recognition ranges from simple detection to various levels of classi-

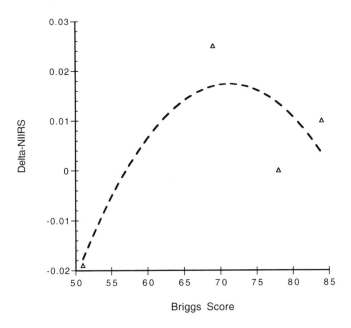

**Figure 5.19** NIIRS/Briggs data for high-resolution displays.

fication and identification. Detection may require a visual search of a display field or detection of a change or difference in signal strength.

Once an image is seen by an observer, some degree of learning takes place. Thus, a large number of independent image or display samples are needed in order to collect sufficient data unbiased from learning effects. With artificial targets, this is not a problem. For example, the observer may be asked to determine in which of four possible locations a target, such as a low-contrast square, appears. The same target can be used many times so long as location is randomized among the possible locations and a correction for guessing is applied to results. With real images, learning becomes a problem. Each image is unique, and once a target has been seen in a particular image, its presence and location or identification may be remembered. This appears to be more of a problem in the surveillance and reconnaissance (S&R) field as opposed to the medical field. S&R imagery shows a wide range of acquisition parameters, whereas medical imagery of a particular type—e.g., lung radiographs—is more standardized in terms of acquisition parameters. This generally provides the opportunity for a larger homogeneous sample for use in performance studies.

The performance of an observer varies as a function of the observer's training and experience as well as the mindset of the observer. In a binary choice situation (two-alternative forced choice), there are four possible outcomes (Fig. 5.20). Not all of the outcomes may be perceived as having equal value. The cost of a miss in an initial medical screening exam, for example, may be perceived as having a greater cost than a false alarm. Depending on factors such as the relative perceived cost of a miss vs. a false alarm, and the perceived ratio of target to nontarget trials,

PERCEPTUAL QUALITY AND UTILITY MEASURES 99

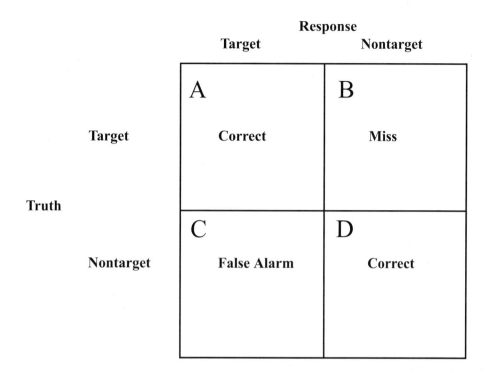

**Figure 5.20** Binary choice outcomes.

the observer's detection threshold will change. The observer may become cautious in reporting a detection (and thus miss targets) or may be aggressive and report targets at the expense of a large number of false alarms.

### 5.5.1 Theory of signal detection

A body of literature called the theory of signal detection (TSD) has been developed to acquire and analyze such decision-making data.[19–22] This theory, commonly used in medical studies, hypothesizes that a display can be characterized in terms of two event distributions: signal (or target) and signal plus noise. Stated another way, some targets are obvious and some are difficult to separate from noise (nontargets). Similarly, noise is sometimes weak and obviously not confused for a target, and at other times may approach the strength of a weak signal. The two distributions are plotted as shown in Fig. 5.21. The degree of separation between the two distributions is defined as the parameter $d'$. The decision criterion $\beta$ defines the observer's threshold in distinguishing a target or signal from a nontarget or noise. The conservative observer reports only the obvious (strong signal) targets, and that observer's $\beta$ moves to the right. The $\beta$ of the less conservative observer moves to the left. The letters A through D in Fig. 5.21 indicate the decision outcome as shown in Fig. 5.20.

TSD data are typically collected by having multiple observers rate their confidence in their decisions regarding a series of observations. The observers are shown a series of events or stimuli and asked to determine whether or not a signal is present and how much confidence they have in their decision. Historically, five category scales (5 is very confident, 1 is weakly confident) have commonly been used. The confidence ratings are used to assess the observer's β. The successful use of confidence ratings requires that observers distribute their ratings across the scale, which does not always happen as desired. The use of 100-point scales has been proposed as a means of overcoming this problem.[22]

A common method of presenting TSD data is a receiver operating characteristic or ROC (Fig. 5.22). The probability of detection (correct response to signal

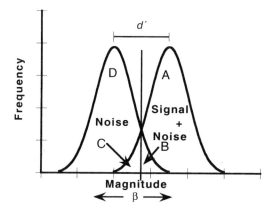

**Figure 5.21** Illustration of signal detection relationships.

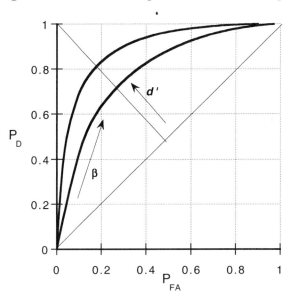

**Figure 5.22** Receiver operating characteristic (ROC) showing probability of detection ($P_D$) vs. probability of false alarm ($P_{FA}$). Reprinted with permission from *Surveillance and Reconnaissance Imaging Systems: Modeling and Performance Prediction*, by Jon C. Leachtenauer and Ronald G. Driggers, Artech House Inc., Norwood, MA, USA (www.artechhouse.com).

**Figure 5.23** Free-response operating characteristics (FROCs). Reprinted with permission from *Surveillance and Reconnaissance Imaging Systems: Modeling and Performance Prediction*, by Jon C. Leachtenauer and Ronald G. Driggers, Artech House Inc., Norwood, MA, USA (www.artechhouse.com).

present) is plotted against the probability of false alarms (incorrect response to noise). A single curve plots the results for a given task and level of display quality as the observer's decision threshold ($\beta$) changes. The conservative observer will have few false alarms but will also have few correct detections. As $\beta$ moves to the left (Fig. 5.21), the number of correct detections and false alarms increase. As the quality of the display improves, the ROC moves away from the negative diagonal. The distance between the inflection point and the negative diagonal is termed d' and is an indication of the relative quality of the presentation, i.e., the distinction between the signal and signal plus noise distributions. The area under the ROC is sometimes used as a measure of performance.

Although ROCs provide an accurate representation of performance that allows comparisons to be made across studies, they require a substantial amount of data for their construction. It is thus not uncommon to find studies in the literature where observed differences between two treatments are found to be not statistically significant.

Two other methods related to ROC analysis have also been developed. An extension of the approach to cases where multiple decisions may be made (multiple targets present) has been developed. Data are plotted as a free-response operating characteristic or FROC (Fig. 5.23).[23]

Where two treatments must be compared, each can be tested separately by creating an ROC for each and comparing them. Alternatively, decisions and confidence ratings can be made in terms of the difference between two treatments. This is called the differential receiver operating characteristic or DROC.[23]

*5.5.2 Time measures*

The time required to perform a task is sometimes used as a measure of quality. Typically, image observers have little time to make what are often complex judgements. This is particularly true for a search task, whether it be a radiologist or a military observer performing the search. The time required to perform both the total task as well as elements of the task is thus of interest.

The search process involves scanning an image to find one or more objects of interest. The scanning process entails making successive fixations (in essence, staring at a single point on the image) separated by rapid interfixation jumps called saccades. It is possible to measure fixation durations and locations and to thus define relationships with quality. Such techniques are also used to study the relationship between duration and accuracy of response. Search studies have shown that the search pattern is not uniform and often incomplete;[24, 25] often the target of interest is fixated but not reported.[26] Finally, search patterns vary among individuals and as a function of experience.[27, 28]

## 5.6 Summary

Perceptual measures may assess either quality or utility. Perceptual quality measures include ranking and rating scales, and ratings of specially constructed targets that measure resolution or contrast discrimination. The Briggs target is the most useful of these targets for display evaluation because it measures spatial discriminability over a range of contrast and luminance values.

Perceptual utility measures involve both the visual and cognitive systems to estimate (subjective) or measure (objective) the display's ability to perform a defined task. The NIIRS is a subjective task performance measure that has been shown to relate closely to Briggs ratings.

Objective performance measurement is commonly used in the medical community. Theory of signal detection (TSD) methods are used to generate receiver operating characteristics (ROCs) to compare detection to false alarm rates. Although ROCs represent an accurate representation of performance that allows comparisons to be made across studies, they require a substantial amount of data for their construction. An extension of the ROC approach to cases where multiple decisions may be made has also been developed (FROC), as has a method to directly compare two treatments (DROC). Finally, measures of visual fixation times and patterns are a useful tool in studying the effects of display quality.

## References

[1] J. C. Leachtenauer and R. G. Driggers, *Surveillance and Reconnaissance Imaging Systems*: *Modeling and Performance Prediction*, Artech House, Boston, MA (2001).

[2] J. C. Leachtenauer, "National Imagery Interpretability Rating Scales: overview and product description," *ASPRS/ASCM Annual Convention and Exhibition Technical Papers, Vol. 1, Remote Sensing and Photogrammetry*, American Society for Photogrammetry and Remote Sensing and American Congress on Surveying and Mapping, Baltimore, MD, April 22–25, 1996, pp. 262–271.

[3] R. Janssen, *Computational Image Quality*, SPIE Press, Bellingham, WA (2001).

[4] International Telegraphy Union, *Methodology for the Subjective Assessment of the Quality of Television Pictures*, Recommendation ITU-R BT.500-10, ITU, Zurich (1974–2000).

[5] R. B. Voas, W. Frizzell, and W. Zink, *Evaluation of Methods for Studies Using the Psychophysical Theory of Signal Detection*, Environmental Research Institute of Michigan, Washington, DC (1987).

[6] M. G. Samet, *Checker Confidence as Affected by Performance of Initial Image Interpreter*, Technical Research Note 214, U.S. Army Behavioral Science Research Laboratory, Washington, DC (1969).

[7] J. M. Irvine and J. C. Leachtenauer, "A methodology for developing imagery interpretability rating scales," *ASPRS/ASCM Annual Convention and Exhibition Technical Papers, Vol. 1, Remote Sensing and Photogrammetry*, American Society for Photogrammetry and Remote Sensing and American Congress on Surveying and Mapping, Baltimore, MD, April 22–25, 1996, pp. 273–279.

[8] J. Leachtenauer, W. Malila, J. Irvine, L. Colburn, and N. Salvaggio. "General Image Quality Equations: GIQE," *Applied Optics*, Vol. 36(32), pp. 8322–8328 (1997).

[9] K. Reihl and L. Maver, "A comparison of two common image quality measures," *Proc. SPIE*, Vol. 2829, pp. 242–254 (1996).

[10] L. M. Maver et al., "Aerial imaging systems," in J. Sturge, V. Walworth, and A. Shepp (eds.), *Imaging Processes and Materials*, Neblettes Eighth Edition, Van Nostrand Reinhold, New York (1989).

[11] E. L. Gliatti, "Image evaluation methods," in *Proc. SPIE*, Vol. 136, pp. 6–12 (1978).

[12] S. J. Briggs, *Digital Display Test Target Development*, D180-15860-1, Boeing Aerospace Co., Seattle, WA (1977).

[13] S. J. Briggs, *Manual Digital Test Target BTP#4*, D180-25066-1, Boeing Aerospace Co., Seattle, WA (1979).

[14] J. Leachtenauer and N. Salvaggio, "Color monitor calibration for display of multispectral imagery," Society for Information Display, International Symposium, *Digest of Technical Papers*, Boston, MA, May 13–15, 1997, pp. 1037–1040.

[15] R. Van Metter, B. Zhao, and K. Kohm, "The sensitivity of visual targets for display quality assessment," *Proc. SPIE*, Vol. 3658, pp. 254–268 (1999).

[16] J. C. Leachtenauer and N. L. Salvaggio, "NIIRS prediction: use of the Briggs target," *ASPRS/ASCM Annual Convention and Exhibition Technical Papers, Vol. 1, Remote Sensing and Photogrammetry*, American Society for Photo-

grammetry and Remote Sensing and American Congress on Surveying and Mapping, Baltimore, MD, April 22–25, 1996, pp. 282–291.

[17] J. Leachtenauer, A. Biache, and G. Garney, "Effects of ambient lighting and monitor calibration on softcopy image interpretability," *Final Program and Proceedings, PICS Conference*, Society for Imaging Science and Technology, Savannah, GA, April 25–28, 1999, pp. 179–183.

[18] J. Leachtenauer, A. Biache, and G. Garney, " Effects of pixel density on softcopy image interpretability," *Final Program and Proceedings, PICS Conference*, Society for Imaging Science and Technology, Savannah, GA, April 25–28, 1999, pp. 184–188.

[19] J. Leachtenauer, D. Griffith, and J. Irvine, *Commercial Analyst Workstation Characterization Study*, National Exploitation Laboratory, Washington, DC, (1992).

[20] J. A. Swets (ed.), *Signal Detection and Recognition by Human Observers*, John Wiley & Sons, Inc., New York (1964).

[21] D. M. Green, *Signal Detection Theory and Psychophysics*, Robert Krieger Publishing Company, Huntington, NY (1974).

[22] J. P. Egan, *Signal Detection Theory and ROC Analysis*, Academic Press Inc., New York (1975).

[23] C. E. Metz, "Fundamental ROC analysis," in J. Beutel, H. L. Kundel, and R. L. Van Metter (eds.), *Handbook of Medical Imaging, Vol. 1, Physics and Psychophysics*, SPIE Press, Bellingham, WA, pp. 751–770 (2001).

[24] D. C. Chakraborty, " The FROC, AFROC, and DROC versions of the ROC analysis," in J. Beutel, H. L. Kundel, and R. L. Van Metter (eds.), *Handbook of Medical Imaging, Vol. 1, Physics and Psychophysics*, SPIE Press, Bellingham, WA, pp. 771–796 (2001).

[25] W. J. Tuddenham and W. P. Calvert, "Visual search patterns in roentgen diagnosis," *Radiology*, Vol. 76, pp. 255–256 (1961).

[26] J. M. Enoch, *The Effect of the Size of the Display on Visual Search*, Mapping and Charting Research Laboratory, Ohio State University, Columbus, OH (1958).

[27] H. L. Kundel, "Visual search in medical images," in J. Beutel, H. L. Kundel, and R. L. Van Metter (eds.), *Handbook of Medical Imaging, Vol. 1, Physics and Psychophysics*, SPIE Press, Bellingham, WA, pp. 837–858 (2001).

[28] H. L. Kundel and P. La Follette, "Visual search patterns and experience with radiological images," *Radiology*, Vol. 103, pp. 523–528 (1973).

[29] M. Kalk and J. M. Enoch, *An Objective Test Procedure Designed to Aid in Selecting, Training, and Rating Photointerpreters*, Mapping and Charting Laboratory, Ohio State University, Columbus, OH (1958).

# Chapter 6
# Performance of the Human Visual System

This chapter will discuss the performance of the human visual system (HVS). The discussion begins with a brief description of the visual system, followed by performance data on the HVS. Performance is discussed in terms of the observer's ability to see small and low-contrast image detail, as well as the ability to detect and describe color differences.

Individuals differ in terms of their visual abilities, and abilities change as people age. Knowledge of these differences and changes is important for optimizing a display. Theoretical and empirical models designed to predict the performance of the HVS can optimize the setup of displays, and are described here in that context.

## 6.1 Physiology of the HVS [1-4]

The HVS can be compared to a camera's operation (Fig. 6.1). Light passes through the cornea and pupil to the lens. The cornea acts as a preliminary focusing mechanism and may also act as a light filter if injury or aging of the cornea has occurred. The iris adjusts the size of the pupil to admit more or less light depending on the light intensity. If the intensity is low, the area of the pupil is enlarged to capture more light. If the light is bright, the pupil contracts. The iris and pupil serve a function similar to the aperture of a camera. However, it takes some time for the pupil size to respond to changes in intensity, a process called adaptation. When we move from a bright room to a dark one, we need some time to adapt in order to see well in the dark room. Various drugs may also affect the size of the pupil and the time required for adaptation. A much greater portion of adaptation takes place at the retina. If we consider the retina as analogous to film in a camera, the effect is similar to changing film sensitivity. When moved from a bright to a dark environment, the pupil adapts in a matter of several seconds while the retina needs several minutes.

The change in the diameter of the pupil as a function of scene luminance is shown in Fig. 6.2. A change in the pupil diameter affects both the angular extent of the light ray bundle entering the eye and the illuminance on the retina. A measure of illuminance called the troland (Td) is used to measure the light falling on the retina. Retinal illuminance is defined as

$$E = A \times L \times (1 - d^2 + 0.0000419 d^4), \qquad (6.1)$$

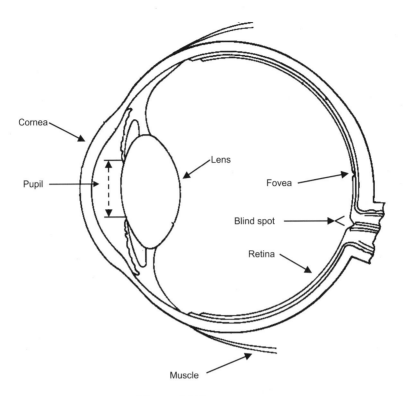

**Figure 6.1** Eye structure.

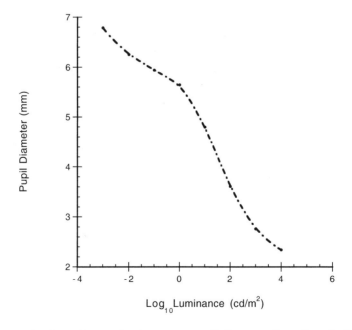

**Figure 6.2** Effect of scene luminance on pupil diameter. Data from Ref. [2].

where $E$ is the retinal illuminance in Td, $A$ is the area of the pupil (measured in mm$^2$), $L$ is the scene luminance (measured in cd/m$^2$), and $d$ is the diameter of the pupil in mm. The function involving $d$ is a correction for the Stiles-Crawford effect, which describes the effect of decreasing brightness (and efficiency) of a light ray as the point of entry gets farther away from the center of the pupil.

As light passes through the lens, the lens attempts to focus the light onto the cornea. Again, the process is not instantaneous. As the distance between the eye and the object we are trying to see changes, the shape of the lens must change in order to properly focus. The process is similar to that of focusing a camera lens. The process of changing focus is called accommodation. As we age, our ability to change the shape of the lens diminishes and we are forced to insert one or more lenses (e.g., bifocals) into the visual chain.

Just as in a camera, when light falls on the retina the image on the retina is inverted. The brain ultimately makes the image look right side up. The retina contains two types of light-sensitive receptors or detectors: cones are concentrated near the center of the retina, and rods are located toward the periphery of the retina (Fig. 6.3). The rods are more sensitive to light but cannot distinguish wavelength (color) differences.

When we stare at a point, our vision is centered on the visual axis (Fig. 6.4). The visual axis passes through the lens to the center of the retina, called the fovea.

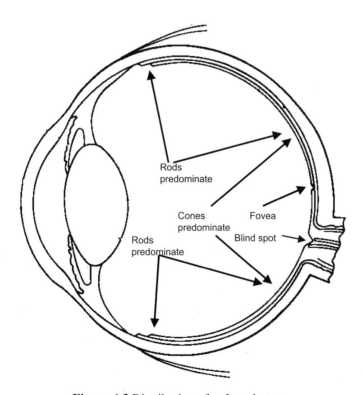

**Figure 6.3** Distribution of rods and cones.

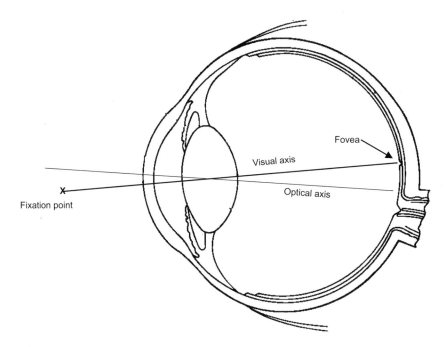

**Figure 6.4** Geometry of the eye.

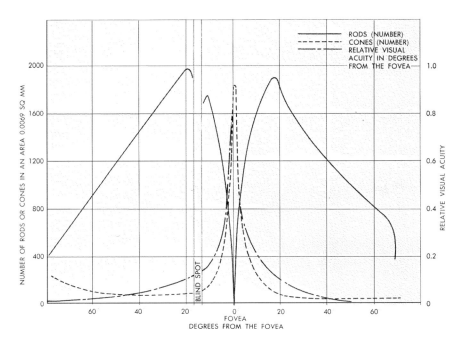

**Figure 6.5** Acuity as a function of rod and cone distribution. Reprinted by permission of University of California Press from *Human Engineering Guide for Equipment Designers*, W. E. Woodson and D. W. Conover. © 1964 The Regents of the University of California.

This is called foveal vision. We also have the ability to see off the visual axis. If you concentrate on a point on this page, you are also aware of objects surrounding the page. This is called peripheral vision. Foveal vision occurs in the central two degrees of vision; peripheral vision occurs outside that region with varying degrees of success.

The rods and cones are connected to nerve cells. Together, they change the light energy to neural impulses that are transmitted to the brain. Near the fovea, each cone is connected to one or more nerve cells. Toward the periphery, several rods and cones may be connected to a single cell. This has the effect of increasing light sensitivity at the expense of resolution (visual acuity). Figure 6.5 shows the distribution of rods and cones along with a measure of relative acuity.

Although foveal vision is used to examine detail in an image, peripheral vision is important in the process of search. A search is accomplished by a series of fixations interspersed with saccades from one fixation point to the next.[5] A fixation, which occurs when we stare at a point using foveal vision, typically lasts for less than half a second. The location of successive fixations is guided in part by what is seen in the periphery. The eye is attracted to bright or high-contrast points in the visual field. Figure 6.6 shows a typical search pattern on an image.[6] The duration of a saccade depends on the angular distance of the saccade, which in turn depends

**Figure 6.6** Search pattern. Reprinted with permission from *Surveillance and Reconnaissance Imaging Systems: Modeling and Performance Prediction*, by Jon C. Leachtenauer and Ronald G. Driggers, Artech House Inc., Norwood, MA, USA (www.artechhouse.com).

on the characteristics of the display and viewing conditions. Limited viewing time and large displays increase the angular distance of saccades. Image blur and other quality degradations decrease saccade distance and duration. A key point is that we do not "see" during a saccade. Thus, as saccadic distances increase, more of the display will not be perceived.

It was noted earlier that only cones are sensitive to color. The trichromatic theory of vision states that there are three types of cones, each sensitive to a different wavelength range roughly corresponding to red, green, and blue light. They are known as the L, M, and S (long, medium, short) wavelength receptors. If we are looking at an object, the color we perceive depends on the wavelength distribution and intensity of illuminating energy, characteristics of the surface we are viewing, and properties of the HVS. It may also depend in part on what we expect to see. The term "color constancy" refers to the extent to which a given object will appear to be the same color under different viewing conditions. The visual system appears able to achieve color constancy over a relatively wide range of viewing conditions.[9]

The discussion thus far has considered only a single eye, or monocular vision. Binocular vision involves both eyes and fuses the two images on the retina to form a single image. Because the eyes are not coincident, we are able to gain a perception of depth due to the angular disparity between objects seen by the two eyes. Objects farther away subtend smaller visual angles (Fig. 6.7) until there is no longer a detectable angular disparity. Angular disparity, along with other monocular visual cues such as perspective, overlap, relative object size, and shadowing, provides a perception of depth. The perception of depth can be simulated with two images obtained from different perspectives using some form of stereoscopic (stereo) viewing. Stereo viewing entails some type of presentation where each eye is presented with a different image, either simultaneously or alternately. Stereo electronic displays use a flicker mechanism, whereby the left and right eyes are alternately presented with the left and right images. The two images are fused to provide the

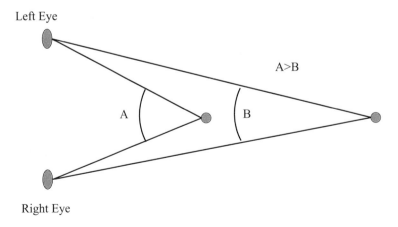

**Figure 6.7** Stereo viewing geometry.

PERFORMANCE OF THE HUMAN VISUAL SYSTEM

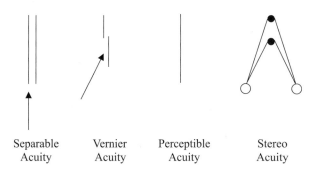

**Figure 6.8** Visual acuity measures.

same depth perception as unaided binocular viewing. The perception of depth can be enhanced by acquiring the two images with wider lateral disparity than normally seen by the eye. Other means of providing a stereo display include polarization (polarized light and polarized glasses such that each eye sees only one image) and color (red for one image/eye and blue for the other).

## 6.2 Visual Performance

Visual performance is measured in terms of the ability to see small detail, low contrast and luminance detail, and color. Visual acuity is a term used to define our ability to see fine detail. Visual acuity can be both measured and defined in a variety of ways (Fig. 6.8). For example, separable acuity measures the ability to resolve or separate closely spaced detail; perceptible acuity measures the ability to see single points or lines; vernier acuity measures the ability to see small lateral displacements along a line; and stereo acuity defines the smallest lateral disparity providing a height or depth difference determination.

Acuity varies as a function of factors other than detail size. The predominant factor is contrast. As contrast decreases, the ability to see small detail also decreases. Because visual acuity and contrast sensitivity are so closely related, they are discussed together. Other factors affecting acuity and sensitivity include the presence of noise, image motion, and viewing time. The discussion that follows addresses minimum separable acuity and briefly reviews the issue of stereo acuity.

### *6.2.1 Separable acuity*

The relationship between separable acuity and contrast can be defined in terms of the "J" curve.[2] The J curve (Fig. 6.9) defines the relationship between the ability to resolve closely spaced detail (minimum separable acuity) and contrast. Acuity is defined in terms of the ability to separate periodic bar patterns or sine wave patterns (measured in cy/deg). Contrast is measured in terms of Cm. Contrast modulation values above the curve are detectable by the average observer, while those

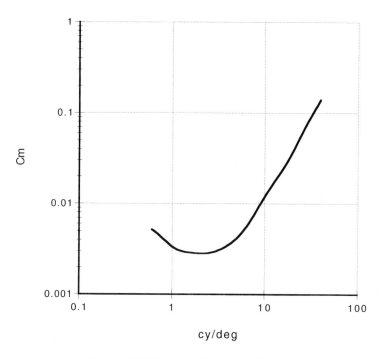

**Figure 6.9** "J" curve. Data from Ref. [2].

below the curve are not. The figure indicates that the ability to detect fine detail (40 cy/deg) requires that Cm be at 0.2 or higher. The data assume that display luminance is at least 15 fL. In reference to Eq. (4.11), the equation for Cm, if the dark bar is at 15 fL, the light bar must be at ~31 fL in order to resolve 40 cy/deg. As Cm decreases, so does acuity. At a Cm of 0.01, no more than ~10 cy/deg can be resolved.

The data shown in Fig. 6.9 can also be interpreted in the context of contrast sensitivity. Where the detail is at 20 cy/deg, Cm must be at 0.07 or higher. If the dark bar is at 15 fL, the light bar can be at ~17 fL. Maximum contrast sensitivity is achieved at 2 cy/deg.

The same data can be expressed in terms of a typical monitor. Assuming two 21-in. CRTs with pixel densities of 100 and 200 ppi viewed at a nominal 18-in. distance, the data in Fig. 6.9 can be relabeled as shown in Fig. 6.10. With no magnification, Cm values less than roughly 0.02 cannot be seen at maximum resolution on the 100-ppi monitor and Cm must be nearly at 0.1 to be seen on the 200-ppi monitor. In order to achieve maximum contrast sensitivity on either monitor, the image must be magnified such that effective pixel density is decreased (moves to the left). Without magnification, the visual Cm threshold of less than 0.003 cannot be achieved.

The data shown in Figs. 6.9 and 6.10 were drawn from a number of studies. Another way of portraying the information is in terms of the MTF of the HVS.[7] The MTF defines the loss in contrast sensitivity as a function of spatial frequency.

PERFORMANCE OF THE HUMAN VISUAL SYSTEM

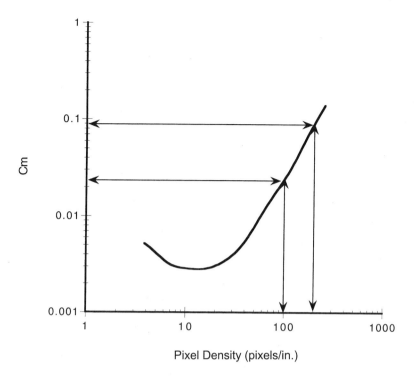

**Figure 6.10** Effect of monitor addressability on Cm detection.

An example is shown in Fig. 6.11. The MTF of the eye is expressed as a Gaussian function defined as

$$M_{opt}(u) = e^{-2\pi^2\sigma^2 u^2}. \qquad (6.2)$$

The value σ is the standard deviation of the line spread function convolved with all of the elements of the visual chain. The value $u$ is the spatial frequency expressed in angular units for the eye. As was the case for the J curve, the vertical extent—and to a lesser degree, the shape of the function—differ with luminance and noise. These relationships will be described in more detail in Sec. 6.4.

Given the MTF, the modulation threshold can be defined at a given spatial frequency. The modulation threshold ($m_t$) is the smallest signal that can be detected above noise. Detection thresholds are defined by TSD methods (Sec. 5.5.1). The inverse of the modulation threshold defines contrast sensitivity:

$$S = \frac{1}{m_t}. \qquad (6.3)$$

As the modulation threshold becomes very small, the sensitivity value increases. Figure 6.12 shows sensitivity data as a function of luminance.[8] Equations describ-

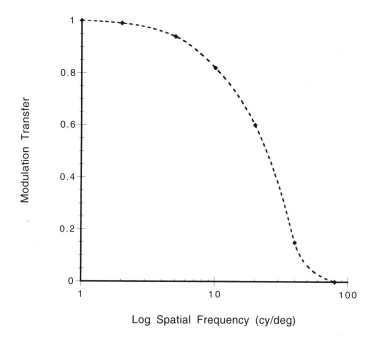

**Figure 6.11** Modulation transfer function (MTF).

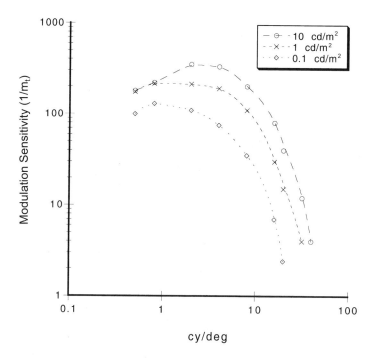

**Figure 6.12** Contrast sensitivity vs. luminance. Data from Ref. [8] used with permission of Elsevier.

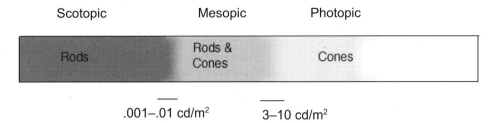

**Figure 6.13** Luminance and vision type.

ing the relationship between Cm sensitivity and spatial frequency are called contrast sensitivity functions or CSF. Increasing luminance increases the values of sensitivity and also appears to move the peak sensitivity to higher spatial frequency values. Other factors affecting the sensitivity function include noise, field of view, monocular vs. binocular viewing (binocular is better), and target type.

When luminance is decreased beyond a certain point, there is insufficient luminance for the cones to respond and thus the rods take over. When luminance is sufficient for the cones to predominate in vision, vision is said to be daylight or photopic. As luminance decreases, vision shifts to the mesopic range; both rods and cones operate in this range. The role of the cones diminishes as luminance decreases to the point where only the rods operate—a range called night time or scotopic vision. Figure 6.13 shows the relationship between types of vision and luminance levels. The dividing line between mesopic and scotopic vision is in the range of 0.001 to 0.01 cd/m$^2$, and the division between mesopic and photopic is in the range of 3–10 cd/m$^2$. Variations in measurement conditions produce the range. The HVS has a total DR of about 90 dB, extending from ~0.0003 cd/m$^2$ to 0.01 cd/m$^2$ for the rods and ~0.01 cd/m$^2$ to 32,000 cd/m$^2$ for the cones.[9]

In the photopic region, the eye is most sensitive to wavelengths in the green/yellow region, as shown in Fig. 6.14. In scotopic vision, maximum sensitivity shifts toward the blue, with slightly less sensitivity in the red. Sensitivity in the mesopic region falls between that shown for the photopic and scotopic.

The ability to resolve detail also varies with motion. Two types of motion are important. As one example, a moving car can be tracked with a camera to "freeze" the motion of the car. However, the background will be blurred. Motion can also be tracked visually. So long as the motion is smooth and continuous, the HVS suffers no loss of acuity up to rates of ~30 deg per second.[2] On a 21-in. display, this is roughly equivalent to following an object from the top to the bottom of the display in 1 second. Note that when we scroll through pages of text in a word processing program, the motion is not smooth and continuous and we generally cannot read the text until we stop scrolling.

An observer may also fixate a point and scroll the image past that point. Here, acuity is degraded at much lower rates. A 50% loss in acuity can occur at a rate of 10 deg/sec.[10] Figure 6.15 shows the relationship between image motion and acuity. On a 21-in. display, a 10-deg/sec rate allows 3.5 sec for the display to be vertically scrolled past the observer.

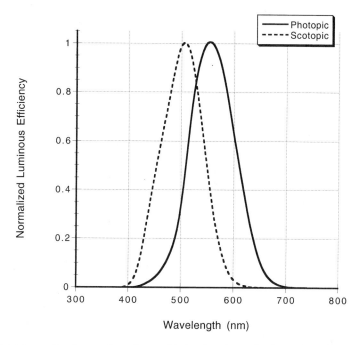

**Figure 6.14** Normalized wavelength sensitivity in photopic (cones) and scotopic (rods) vision. Data from Ref. [3] based on the CIE photopic and scotopic sensitivity functions.

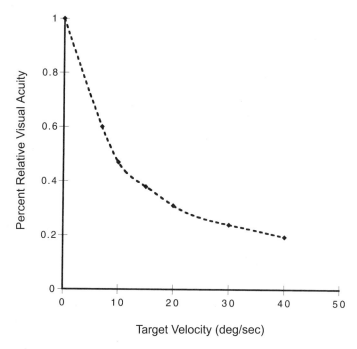

**Figure 6.15** Effect of image motion on visual acuity. Data from Ref. [10] used with permission of Elsevier.

PERFORMANCE OF THE HUMAN VISUAL SYSTEM 117

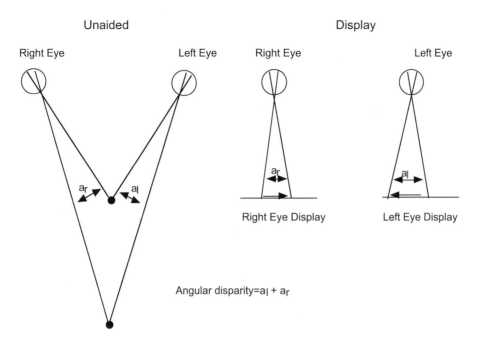

**Figure 6.16** Measurement of lateral disparity.

## 6.2.2 Stereo acuity

Stereo acuity is measured in terms of the minimum detectable lateral disparity. Lateral disparity is the horizontal angular disparity difference. Figure 6.16 shows the measurement for both unaided vision and for a display. Although thresholds as low as 0.1 arc minute have been measured in studies, more realistic viewing conditions produce higher thresholds. Figure 6.17 shows one set of data.[11] As is the case with minimum separable acuity, low contrast degrades performance. As the distance between two objects to be compared increases, the threshold increases.

## 6.2.3 Color vision performance

The trichromatic theory of color vision was discussed previously in Sec. 6.1. When luminance is sufficient for photopic vision, we see color. Under ideal circumstances, we are able to distinguish very small color differences. Differences of 0.004 in $u'v'$ space and 1 $\Delta C$ unit (chroma) are said to be just distinguishable.[12] These values, however, represent the ability to match color under perfect viewing conditions. Such conditions require that the two colored areas to be matched are of equal size and equal luminance, and have equal luminance and color of their surrounds. Further, such values do not hold over the full color space as indicated by the MacAdam ellipses shown in Fig. 2.12.

Figure 6.18 shows $\Delta C$ values as a function of changes in viewing conditions. The values shown represent changes in perceived hue and/or saturation as a func-

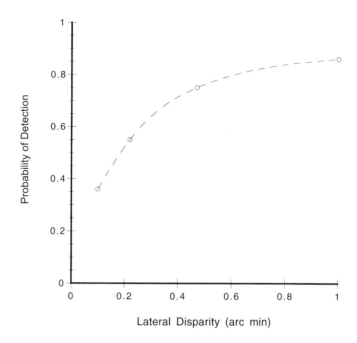

**Figure 6.17** Probability of detecting a raised disc in an array of discs varying in size and contrast. Data from Ref. [11].

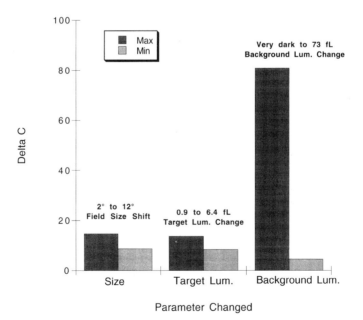

**Figure 6.18** Effect of parameter changes on ΔC change. Data from Ref. [2].

Table 6.1 Effects of viewing condition changes.

| Parameter Change | Hue | Saturation |
| --- | --- | --- |
| Size: 2 deg to 12 deg | Shift away from reference surround | Increased |
| Target luminance: 0.9 to 6.4 fL | Shift away from reference surround | Increased |
| Background luminance: very dark to 73 fL | Little change | Increased |

tion of differences between the two colors being compared. Each pair of bars represents the minimum and maximum ΔC values observed over a large gamut. The effect of each change varied depending on the location in color space. Perhaps as important as the magnitude of the changes is the direction of change. Table 6.1 summarizes the changes in hue and chroma as a function of viewing condition changes, which are defined in terms of the reference surround.

The data shown in Fig. 6.18 suggest the difficulties in matching colors on most imagery. In most real imagery, colored areas to be matched are not of the same size and do not have the same surround. The problem is even greater when one considers that luminance and chromaticity are not constant across the face of a display. Thus, even if two areas have the same chromaticity, they are unlikely to appear the same because of display luminance or chromaticity differences. Differences of 5 to 8 ΔC across a display are not uncommon.

The data also indicate possible ways of improving color matching or discrimination performance. The two areas of interest should be centered in the display and should be somehow masked to have the same area and background. A far simpler approach, however, is to use a software program that defines the color coordinates of the two areas of interest.

## 6.3 Individual Differences

Thus far, the visual performance data shown have been the average of several observers, or more often, that of one or two observers. There are in fact substantial differences among people in terms of visual performance. Figure 6.19 shows the range in Cm over a group of observers as a function of the target size (angular subtense).[13] Observers were required to detect a small square target and to detect a break in a Landolt C (see Fig. 5.7). Background luminance was set at 1.2 fL.

Figure 6.20 shows the variation in performance for the Briggs C-7 target readings (average across eight targets) on a monochrome monitor.[14] The average score was 69 and the standard deviation was 11. Figure 6.21 shows variability for the color Briggs target.[15] The average was 58.3 and the standard deviation was 6.7.

There are a variety of reasons why visual performance differs among observers. First, as was discussed in Sec. 5.5, detection thresholds may differ. The more conservative observers may report only those targets they are absolutely certain they can detect or resolve.

Second, a variety of possible vision problems can affect performance. The lens of the eye may not be able to correctly focus light rays on the retina. Near-sighted

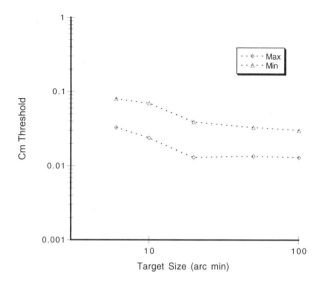

**Figure 6.19** Cm variability. Data from Ref. [13] used with permission of the Optical Society of America.

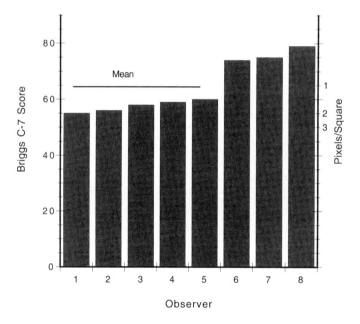

**Figure 6.20** Observer variability in Briggs scores.

people (myopes) focus distant objects in front of the retina. The objects are thus out of focus unless a corrective lens is applied. However, myopes are often able to focus at very close distances, thus improving their ability to resolve detail by increasing the visual angle subtended by the detail (providing that they remove

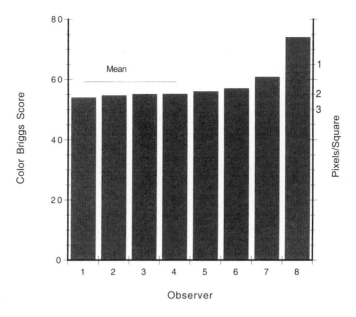

**Figure 6.21** Observer variability in color Briggs scores.

their glasses). Far-sighted people (presbyopes) focus near objects behind the retina. This means that they need to increase the minimum viewing distance in order to correctly focus. This is somewhat counterproductive, however, in that the visual angle subtended by the material viewed is decreasing. Individuals with astigmatism focus light in different orientations in different planes. Again, a lens correction is possible.

Focusing of the eye requires a change in the shape of the lens. As we age, our ability to make this change decreases, hence the common need for corrective lenses. However, even in the absence of proper correction, the visual/cognitive system may successfully compensate to some degree. Even though the normal reading distance is 12–14 in., it is possible to move closer to a display and still read (or see detail). This, too, can account for differences in visual performance.

In addition to a general hardening of the eye lenses as we age, other factors degrade vision. The density of the lens increases and some cone sensitivity is lost with age. Figure 6.22 shows the changes in Cm occurring with age in one study.[16] Wavefront aberrations (not corrected with normal glasses) also increase with age, thus degrading acuity. The effect of correcting for these aberrations is shown in Fig. 6.23.[17] In one case, only defocus and astigmatism have been corrected (DF+A); in the other, both monochromatic and chromatic wavefront aberrations have been corrected. Substantial improvement is shown. Color discrimination also degrades with age,[18] and age effects have been seen in Briggs target readings (as shown in Fig. 6.24).[19]

There are individual differences in contrast sensitivity as well as in other visual performance measurements. One study showed stereo acuity (minimum detectable

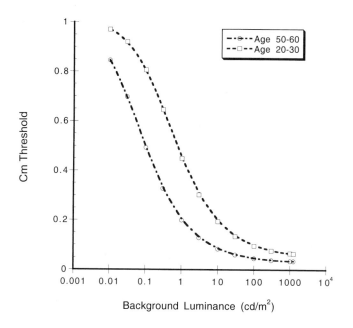

**Figure 6.22** Cm threshold vs. age. Data from Ref. [16] used with permission of the Illuminating Engineering Society of North America.

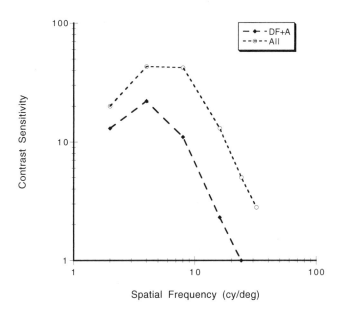

**Figure 6.23** Effect of correcting wavefront aberrations on MTF. Reprinted with modification from Ref. [17] by permission of the Optical Society of America.

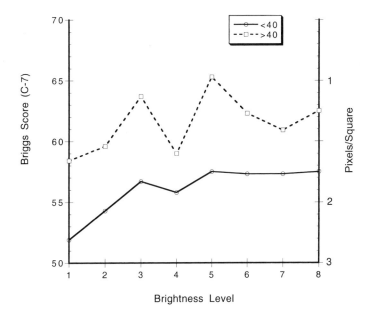

**Figure 6.24** Effect of age on Briggs target scores.

lateral disparity) to range from 0.03 to 1.7 arc minutes.[11] Because peripheral vision is important in search, it has been studied rather extensively. It is clear that acuity degrades as we move away from the fovea. One reason is focus, and improvements in peripheral acuity have been shown with corrective lenses.[20] As is the case with the correction of wavefront aberrations, however, the correction is not entirely practical to implement. Studies on the effects of training to improve peripheral acuity have been inconclusive.[21, 22] It is not clear whether improvements come from paying more attention to the periphery or in learning to deal with out-of-focus images. Finally, studies have shown a positive relationship between peripheral acuity and image search performance.[23, 24]

Some portion of the population suffers from color vision deficiencies. This so-called "color blindness" is actually defined by a difference in the mix of red and green light needed to produce yellow. Individuals whose receptors are deficient in detecting red or green (~6% of the male population) need more than the normal amount of red or green light. Dichromats—individuals who see only two of the three primary colors—constitute about 2% of the male population. They are most commonly "blind" to red or green and cannot see their "blind" color as other than a gray. A small portion (0.002%) of the male population cannot discriminate yellow and blue. Less than 0.5% of the female population suffers from any type of color deficiency compared to 8% of the male population.[2] The tiny portion of the population who cannot see any colors are said to have monochromatic vision. They typically also suffer from other visual defects.[2]

With the exception of major focusing issues that are correctable with spectacles, age-related vision loss is not generally correctable optically. Vision loss can be accommodated in electronic displays to varying degrees, however, with software processing. Techniques such as contrast stretching can overcome contrast losses, and magnification can help to reduce the effects of acuity loss.

## 6.4 Models of Visual Performance

Several models have been developed to predict visual performance as a function of the viewing situation. In the context of displays, these models have potential value in optimizing the luminance and color output function of the display, as well as providing an understanding of the relative importance of physical display quality measures. For example, contrast sensitivity of the eye varies as a function of luminance level. This variation does not parallel the CL/luminance function of displays. The detectability of CL differences is thus not constant across the luminance range of the display. By employing one of the models of contrast sensitivity, a LUT can be generated that makes CL differences equal in terms of contrast sensitivity. Models are described in this section, and their application is discussed in Chapter 10.

### 6.4.1 Monochrome (luminance) models

Early attempts at developing luminance contrast discrimination models were based on empirical studies. Observers were typically asked to perform detection tasks (detect a small target, detect sine wave modulation) as a function of display variables such as target luminance and contrast, surround luminance, and target size and spatial frequency.

One of the earliest models is that developed by R. Blackwell during World War II.[25] Observers were required to detect a spot of projected light on a background of a different luminance. Surround luminance also was varied. Some 450,000 observations were made with varying levels of target size and contrast and surround luminance. Figure 6.25 summarizes some of the data collected when the disc was brighter than its background. Contrast is defined as the luminance difference between the target and background. Note that once the photopic region is reached (log adaptation brightness = 0.47 fL), the rate of change initially decreases. The data clearly show that the ability to detect luminance differences improves with increased target size and adaptation (background) brightness.

Using a different Blackwell data set, another model was developed to predict foveal vision performance as a function of target area, background luminance, presentation time, and threshold contrast.[26]

In a study requiring observers to detect modulation changes in sine wave targets, an equation was developed relating modulation thresholds to target size and frequency, background and surround luminance, and various interaction terms.[27] The equation, with all terms statistically significant, was

$$M_t = -0.52522 - 0.51347A - 0.43732C -$$
$$3.33707D + 0.15254A^2 + 1.13097B^2 + \quad (6.4)$$
$$2.3101D^2 - 0.17248AB,$$

where

$M_t = \log_{10}$ modulation threshold,
$A = \log_{10}$ target luminance (fL),
$B = \log_{10}$ surround luminance (fL),
$C = \log_{10}$ target subtense (deg), and
$D = \log_{10}$ target spatial frequency (cy/deg).

The equation accounted for 91% of the variance in the modulation threshold data collected in the study and used to define the equation.

The Image Display and Exploitation (IDEX) system uses a model based on an empirical detection study[28] that required observers to determine the location of a tab projecting from a larger square at the same luminance. The background was made up of 32 pixel squares with various randomly assigned luminance levels. The tab subtended ~6 minutes and the square subtended ~ 4 degrees. The contrast sensitivity function (CSF) was defined as

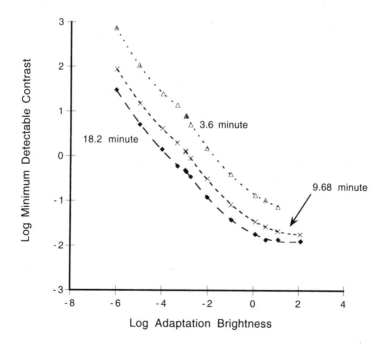

**Figure 6.25** Contrast sensitivity as a function of target size (angular subtense) and surround luminance. Data from Ref. [25] used with permission of the Optical Society of America.

$$d_t = 0.01 + 0.0017L_s + 0.066L, \tag{6.5}$$

where $d_t$ is the minimum detectable luminance difference, $L_s$ is the surround luminance in fL (average of the 32 pixel squares), and $L$ is the lower luminance value (in fL).

Probably the most complex of the contrast sensitivity models is that developed by Barten.[7] The Barten contrast sensitivity model predicts threshold modulation as a function of field size, spatial frequency, and target luminance. It includes the effects of image noise, internal noise (photon and neural), filtering of low spatial frequencies in the ganglion cells (lateral inhibition), the optical MTF of the eye, and target size. The model is currently defined for binocular viewing as

$$\frac{1}{m_t(u)} = \frac{\frac{M_{opt}(u)}{k}}{\sqrt{\frac{2}{T}\left(\frac{1}{x_o^2} + \frac{1}{x_{max}^2} + \frac{u^2}{N_{max}^2}\right)\left(\frac{1}{\eta p E} + \frac{\phi_o}{1 - e^{-\left(\frac{u}{u_o}\right)^2}}\right)}}, \tag{6.6}$$

where $m_t(u)$ is the threshold modulation at frequency $u$, $M_{opt}(u)$ is the optical MTF of the eye at frequency $u$, $k$ is a constant (3.0), and $T$ is the integration time of the eye (0.1 sec).* For monocular viewing, the value 2 under the square root sign is replaced by a 4.

The eye has a limited ability to integrate in the spatial domain. The expression

$$X = \left\{\frac{1}{x_o^2} + \frac{1}{x_{max}^2} + \frac{u^2}{N_{max}^2}\right\}^{-0.5} \tag{6.7}$$

is used to characterize the spatial integration process, where $X_o$ is the angular size of the object (the visual angle subtended by the object), $X_{max}$ is the maximum angular size of the integration area, and $N$ is the maximum number of cycles over which integration can occur. A value of 12 deg is used for $X_{max}$ and 15 cy for $N_{max}$.

The term $1/\eta pE$ defines photon noise, where $\eta$ is total quantum efficiency (0.03) and $p$ is a photon conversion factor (which varies as a function of the light source) to convert light units to units of photon flux density entering the eye(s). For a P4 white phosphor, a value of $1.24 \times 10^6$ per Td[6] is provided for photopic viewing. This constant $p$ times the illuminance $E$ (measured in Td) and the quantum efficiency $\eta$ defines the photon flux passing through the cornea; $E$ is the illuminance of the eye defined as

---

\*   Equations (6.6) through (6.12) are reprinted with permission from *Contrast Sensitivity of the Human Eye and Its Effect on Image Quality*, Vol. PM72, SPIE Press, Bellingham, WA (1999).

$$E = \frac{\pi d^2}{4} L \left[ 1 - \left(\frac{d}{9.7}\right)^2 + \left(\frac{d}{12.4}\right)^4 \right], \tag{6.8}$$

where $d$ is the diameter of the eye pupil (mm) and $L$ is the luminance of the target in cd/m$^2$. Neural noise is defined by $\phi_o$ and has an estimated value of $3 \times 10^{-8}$ sec deg.$^2$ The lateral inhibition process attenuates low spatial frequency components. The MTF of the lateral inhibition process is defined as

$$M_{lat}(u) = \sqrt{1 - e^{-\left(\frac{u}{u_o}\right)^2}}, \tag{6.9}$$

where $u$ is the angular spatial frequency and $u_o$ is the frequency, defined as 7 cy/deg, at which lateral inhibition no longer has an effect. Finally, the optical MTF of the eye is defined as

$$\text{MTF}_{opt}(u) = e^{-\pi^2 \sigma^2 u^2}. \tag{6.10}$$

The term $\sigma$ is the standard deviation of the line spread function due to the convolution of various effects in the eye that contribute to the total effect. It is defined as

$$\sigma = \sqrt{\sigma_o^2 + (C_{ab}d)^2}, \tag{6.11}$$

where $\sigma_o$ is a constant (0.5 arc min), $C_{ab}$ is another constant (0.08 arc min/mm), and $d$ is the pupil diameter.

The model was shown to be a good fit to data from 14 different studies. The studies covered a range of luminance values, spatial frequencies, field sizes, and types of light (monochromatic and white).

The model is capable of accounting for image noise with an added noise term

$$\frac{1}{m_t(u)'} = \frac{1}{\sqrt{m_t(u)^2 + k^2 m_n^2}}, \tag{6.12}$$

where $m_t(u)'$ is the modulation with white noise and $k$ is as previously defined (3.0). The term $m_n$ is the modulation of the noise. It is defined as the average image luminance divided by the luminance SNR. Figure 6.26 shows the effect of noise on contrast sensitivity. The Barten model was developed and validated on conventional CRT displays without specific measurement of display noise characteristics. The noise terms should thus be invoked only when it is desirable to account for image noise characteristics beyond those of the display.

Contrast sensitivity models can be used to create a function showing threshold modulation as a function of luminance. Beginning with some value of Lmin, the first modulation threshold is computed and added to Lmin to make a second

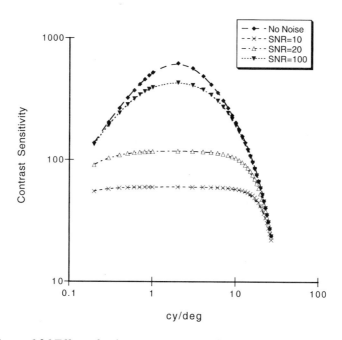

**Figure 6.26** Effect of noise on contrast sensitivity. Data from Ref. [7].

modulation threshold computation. By successively adding and computing modulation thresholds, a function is created with perceptually equal luminance steps. This function is typically different from the normal I/O function of the monitor (Fig. 6.27). A LUT can be defined to transform the native I/O function to the perceptually linearized function.

The original Barten model is used as the basis for the NEMA/DICOM perceptual linearization function. The original function assumes a 2-deg target at 4 cy/deg spatial frequency. This is roughly four times larger than the spatial frequency of a 100-ppi monitor viewed at 18 in. The target size is such that it would occupy 63 pixels on the same monitor. The NEMA/DICOM equation used to define Cm steps (defined as the just-noticeable difference or JND for the defined luminance level) is

$$Log_{10}L_i = \frac{a + c\ln(I) + e[\ln(I)]^2 + g[\ln(I)]^3 + k[\ln(I)]^4}{1 + b\ln(I) + d[\ln(I)]^2 + f[\ln(I)]^3 + h[\ln(I)]^4 + m[\ln(I)]^5}, \quad (6.13)$$

where $a = -1.3011877$,
$b = -0.025840191$,
$c = 0.080242636$,
$d = -0.10320229$,
$e = 0.13646699$,
$f = 0.028745620$,
$g = -0.025468404$,

$h = -0.0031978977$,
$k = 0.00012992634$, and
$m = 0.0013635334$.

The value $I$ is the index or CL value ranging from 1 to 1023, and $L$ is luminance in cd/m². The function can also be approximated by the reversible equation

$$Y = (e^{ax+b} - c)^d, \qquad (6.14)$$

where $Y$ = luminance in fL,
$x$ = JND count variable (1–1023),
$a = 0.00325935271299$,
$b = 0.47562881149555$,
$c = 1.51$, and
$d = 1.87$.

The most recent version of the Barten model differs somewhat from that used in the NEMA/DICOM function shown in Fig. 6.28. The maximum difference is about 8% at a CL of 29 for an 8-bit display. The current version can be defined by

$$Y = (e^{ax^\wedge 2 + bx + c} - d)^e, \qquad (6.15)$$

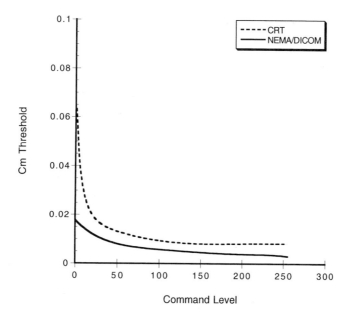

**Figure 6.27** Comparison of perceptually equal (NEMA/DICOM) vs. normal monitor luminance steps.

where $Y$ and $x$ are as in Eq. (6.14):
 a = 0.0000006932,
 b = 0.001432233,
 c = 1.078964690,
 d = –2.863985876, and
 e = 1.707392638.

A key factor in the implementation of the Barten model is the assumption made regarding target size, average luminance, and spatial frequency. Figure 6.29 shows the modulation thresholds as a function of average luminance and spatial frequency. Figure 6.30 shows the effect of target size. The 0.2-deg figure is the angular subtense of a 5-pixel area on a 100-ppi display viewed from 16 in.; the 4-deg size represents a 110-pixel area on a 100-ppi monitor. A 100-ppi monitor viewed from 16 in. has a spatial frequency of 14 cy/deg.

The Barten and NEMA/DICOM models assume conventional imagery in the sense of noncoherent illumination. With coherent illumination, a phenomenon called *speckle* appears in the imagery. Synthetic aperture radar uses coherent illumination, which results in what is called clutter or speckle. Speckle appears when the coherent energy illuminates objects smaller than the illumination wavelength. Figure 6.31 shows an example. It has been shown that the CSF is significantly degraded in the presence of clutter.[30] In a study where two observers were asked to determine the orientation of square wave gratings, the relationship between the CSF with and without speckle was determined. A factor called K defined the ratio

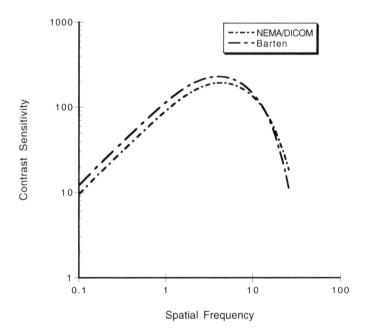

**Figure 6.28** Comparison of Barten models.

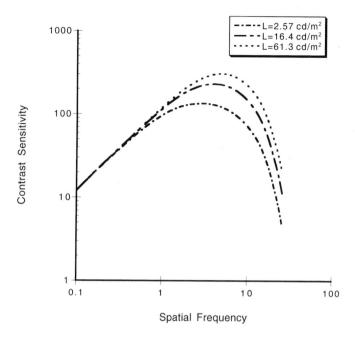

**Figure 6.29** Contrast sensitivity for three luminance levels with target size at 2 deg.

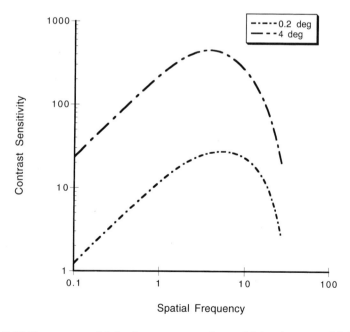

**Figure 6.30** Contrast sensitivity for two target sizes with luminance at 16.4 cd/m$^2$.

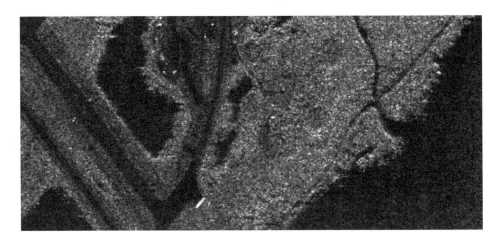

**Figure 6.31** Speckle (clutter) on coherent radar imagery.

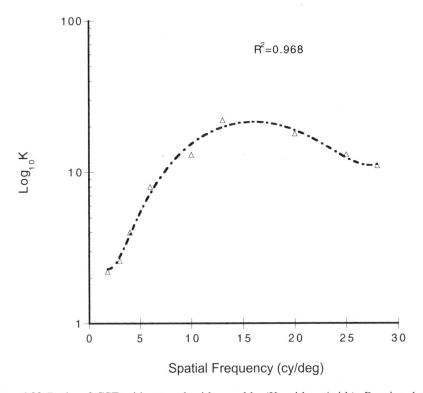

**Figure 6.32** Ratio of CSF without and with speckle (K=without/with). Reprinted with modification from Ref. [30] by permission of the Optical Society of America.

between the CSF with and without speckle. Figure 6.32 shows a function fit to the data. The correction is appropriate with luminance greater than ~ 3 fL. The equation defining the relationship is

$$K = 3.869 - 1.8100 SF + 0.572 SF^2 - 0.033 SF^3 + 0.0006 SF^4, \qquad (6.16)$$

where SF is spatial frequency in cy/deg.

The effect of speckle on the CSF differs from that of noise. The effect was also shown to be independent of luminance level over the spatial frequency range of 4 to 28 cy/deg. The K factor can be applied to the Barten model by first using the Barten model to compute a CSF. Equation (6.16) is used to define the factor K at each spatial frequency and the CSF ($1/M_T$) divided by K to determine the CSF in the presence of speckle.

### *6.4.2 Color models*

For display applications, color models are less well developed than luminance models and are generally less useful. The CIE standard observer data sets and the MacAdam ellipses described in Sec. 2.2 represent empirically derived models. They provide a basis for measuring color and an indication of the ability to discriminate color differences. They are useful if the goal is to reproduce a color image on another display with the same perception of color, but they are not useful in determining whether two colors in a heterogeneous field of colors can be matched or discriminated. Similarly, models describing the color vision process (see, for example, Ref. [3]) are useful in understanding color vision, but again, do not provide the basis for predicting the ability to match or discriminate colors in complex displays.

A variety of models or approaches have been developed to deal with color displays. The CIE $L^*a^*b^*$ and $L^*u^*v^*$ are considered the most useful of several alternatives for display applications.[31] These models provide the basis for color matching and transformation but do not measure detection performance in the same sense as the monochrome models (e.g., Ref. [7]). The models also do not distinguish between luminance and chroma differences in computing ΔE values. A measure called $\Delta E_{IL}$ has been proposed that normalizes luminance differences.[12] Color provides an additional contrast dimension, but a practical application as a method of contrast enhancement has generally not been found.[32] So-called "pseudocolor" schemes, in which closely spaced luminance values are color coded, have been proposed and are sometimes used. However, in order to maximize contrast, it is necessary to apply different hues to closely spaced luminance values. This generally leads to an inability to decipher actual luminance differences. Other schemes that correlate luminance differences with wavelength show less perceptible contrast than do their monochrome counterparts.

For these reasons, this section will not discuss color models further. Additional references will be made in the discussions of display calibration (Chapter 9), pixel processing (Chapter 10), and presentation media (Chapter 11).

## 6.5 Summary

This chapter discussed the operation of the human visual system (HVS) in terms of its receptors. The foveal area of the retina contains cone receptors that are sensitive to color. Because of their dense packing, cone receptors show higher acuity than the surrounding rod receptors. The rod receptors are more sensitive to low light levels and play a key role in the search process as they guide the process of successive fixations.

Performance of the HVS was described in terms of resolution and contrast (both luminance and color). The ability to see fine detail is a function of contrast, and the ability to see small contrast differences is a function of visual subtense or target size. Individual differences, particularly those associated with aging, were described and illustrated. As we age, our ability to see fine detail and low contrast diminishes and is not totally correctable. Therefore, we must use pixel processing to overcome defects of the HVS. Finally, empirically and theoretically derived models of contrast detection were described. Contrast detection thresholds are a function of target size, spatial frequency, and both target and surround luminance. Under the same conditions, the NEMA/DICOM and current Barten models are similar. However, both are sensitive to the assumptions made regarding target size and average scene luminance. These models will be used in subsequent chapters in reference to display calibration, pixel processing, and preparation of presentation media. Although a variety of color models exist, they do not predict color matching or discrimination for typical color imagery where items to be matched or discriminated have different backgrounds, sizes, and brightness levels.

## References

[1]  P. A. Keller, *Electronic Display Measurement*, John Wiley & Sons, Inc., New York (1997).
[2]  R. J. Farrell and J. M. Booth, *Design Handbook for Imagery Interpretation Equipment*, Boeing Aerospace Co., Seattle, WA (1984).
[3]  G. Wyszecki and W. S. Stiles, *Color Science*, Second Edition, John Wiley & Sons, Inc., New York (1982).
[4]  J. W. Wulfeck, A. Weiz, and M. W. Raben, *Vision in Military Aviation*, WADC TR 58-399, Wright Air Development Center, Wright-Patterson Air Force Base, OH (1958).
[5]  A. F. Fuchs, "The neurophysiology of saccades," in R. A. Monty and J. W. Senders (eds.), *Eye Movements and Psychological Processes*, Lawrence Erlbaum Publishers, Hillsdale, NJ (1976).
[6]  G. A. Fry and J. M. Enoch, *Second Interim Technical Report: Human Aspects of Photographic Interpretation*, Ohio State University Mapping and Charting Research Laboratory, Columbus, OH (1956).
[7]  P. J. G. Barten, *Contrast Sensitivity of the Human Eye and Its Effect on Image Quality*, SPIE Press, Bellingham, WA (1999).

[8] A. van Meeteran and J. Vos, "Resolution and contrast sensitivity at low luminances," *Vision Research*, Vol. 12, pp. 825–833 (1972).

[9] W. E. Woodson, B. Tillman, and P. Tillman, *Human Factors Design Handbook,* Second Edition, McGraw-Hill, New York (1992).

[10] B. Brown, "Resolution thresholds for moving targets at the fovea and in the peripheral retina," *Vision Research*, Vol. 12, pp. 293–304 (1972).

[11] R. J. Farrell, C. D. Anderson, and G. P. Boucek, *Construction and Standardization of the CL Stereo Acuity Test (Form 2)*, D180-19059-1,2, The Boeing Company, Seattle, WA (1975).

[12] M. H. Brill and L. D. Silverstein, "Isoluminance color difference metric for application to displays," Society for Information Display, 2002 International Symposium, *Digest of Technical Papers*, Vol. XXXIII, No. 2, pp. 809–811 (2002).

[13] J. J. Vos, A. Lazet, and M. A. Bouman, "Visual contrast thresholds in practical problems," *J. Opt. Soc. Am.*, Vol. 46, pp. 1065–1068 (1956).

[14] J. Leachtenauer, G. Garney, and A. Biache, "Contrast modulation—how much is enough?" Final Program and Proceedings, PICS Conference, Society for Imaging Science and Technology, Portland, OR, March 26–29, 2000, pp. 130–134.

[15] J. C. Leachtenauer and N. L. Salvaggio, "NIIRS prediction: use of the Briggs target," *ASPRS/ASCM Annual Convention and Exhibition Technical Papers, Vol. 1, Remote Sensing and Photogrammetry,* American Society for Photogrammetry and Remote Sensing and American Congress on Surveying and Mapping, Baltimore, MD, April 22–25, 1996, pp. 282–291.

[16] H. R. Blackwell, "Brightness discrimination data for the specification of quality of illumination," *Illum. Engr.*, Vol. 47, pp. 602–609 (1952).

[17] G.-Y. Yoon and D. R. Williams, "Visual performance after correcting the monochromatic and chromatic aberrations of the eye," *J. Opt. Soc. Am. A*, Vol. 19(2), pp. 266–275 (2002).

[18] H. Knau and J. S. Werner, "Senescent changes in parafoveal color appearance: saturation as a function of stimulus area," *J. Opt. Soc. Am. A*, Vol. 19(1), pp. 208–214 (2002).

[19] J. Leachtenauer, G. Garney, and A. Biache, "Effect of monitor calibration on imagery interpretability," *Final Program and Proceedings, PICS Conference*, Society for Imaging Science and Technology, Portland, OR, March 26–29, 2000, pp. 124–129.

[20] M. Millodot, C. A. Johnson, A. Lamont, and H. W. Leibowitz, "Effects of dioptrics on peripheral visual acuity," *Vision Research*, Vol. 15, pp. 1357–1362 (1975).

[21] F. Low, *Effect of Training on Acuity of Peripheral Vision*, PB 50339, Civil Aeronautics Administration, Washington, DC (1946).

[22] C. W. Crannel and J. M. Christensen, *Expansion of the Visual Form Field by Perimeter Training*, WADC-TR-55-368, Wright Air Development Center, Wright Patterson Air Force Base, OH (1958).

[23] R. A. Erickson, "Relation between visual search time and peripheral visual acuity," *Human Factors*, Vol. 6(2), pp. 165–177 (1964).

[24] J. C. Leachtenauer, "Peripheral acuity and photointerpretation performance," *Human Factors*, Vol. 20(5), pp. 537–551 (1978).

[25] R. Blackwell, "Contrast threshold of the human eye," *J. Opt. Soc. Am.*, Vol. 36, pp. 624–643 (1946).

[26] M. Matchko and G. R. Gerhart, "ABCs of foveal vision," *Opt. Eng.*, Vol. 40(12), pp. 2735–2745 (2001).

[27] J. G. Rogers and W. L. Carel, *Development of Design Criteria for Sensor Displays,* Hughes Aircraft Company, Culver City, CA (1973).

[28] C. W. Crites et al., *Visual Tonal Discrimination Characteristics and their Effect on Display Tonal Calibration,* General Electric Corporation, Valley Forge, PA (1986).

[29] National Electrical Manufacturers Association, *Digital Imaging and Communications in Medicine (DICOM), Part 14: Grayscale Standard Display Function*, Rosslyn, VA (2001).

[30] J. M. Artigas, A. Felipe, and M. J. Buades, "Contrast sensitivity of the visual system in speckle imagery," *J. Opt. Soc. Am. A*, Vol. 11(9), pp. 2345–2349 (1994).

[31] H. R. Kang, *Color Technology for Electronic Imaging Devices,* SPIE Press, Bellingham, WA (1997).

[32] E. A. Krupinski, " Practical applications of perceptual research," in J. Beutel, H. L. Kundel, and R. L. Van Metter (eds.), *Handbook of Medical Imaging, Vol. 1, Physics and Psychophysics,* SPIE Press, Bellingham, WA, pp. 895–929 (2001).

# Chapter 7
# Contrast Performance Requirements

This chapter describes certain requirements for optimum display quality including contrast-related luminance and color requirements. The following two chapters treat resolution and noise/artifact requirements. This discussion begins with a listing of key parameters and requirements that were drawn primarily from a set of requirements developed in 1998–1999 by representatives from the National Imagery and Mapping Agency (NIMA) and the National Information Display Laboratory (NIDL). The requirements were developed for the NIMA Integrated Exploitation Capability (IEC) program, a major soft-copy workstation procurement program. These requirements have been modified in some cases by the author to reflect technology improvements as well as findings in the medical literature.

In this chapter, each performance parameter is defined and the effects of performance variations are illustrated with image examples and data from previous studies. Methods of acquiring the information needed to assess the performance of a display are described. Measurement procedures are explained and test targets are provided. These procedures were drawn from several references[1-7] but have been simplified in some cases. Note that all of the test targets are listed in the appendix and have been provided on the enclosed CD.

Methods for determining the performance of a display range from simple visual assessment to complete instrumentation. The choice of method is cost-driven. A complete set of measurement instruments costs on the order of $200,000 U.S. (2002) and can probably be justified only for large procurements by organizations maintaining several hundred displays. At the other end of the spectrum, a reasonable level of performance can be achieved for $1000 or less.

## 7.1 Performance Requirements

Table 7.1 lists key luminance parameters and both minimum and desired performance levels for color monitors. Monochrome requirements are listed in Table 7.2. The requirements for monochrome and color displays are provided separately because of the different levels of performance achieved by the two types of displays and the difference in key parameters. Luminance requirements for stereo display systems are provided in Table 7.3. The stereo requirements are defined at the output of the viewing device, typically some type of glasses. Stereo viewing devices severely limit output luminance.

**Table 7.1** Luminance requirements–Color.

| Measure | Minimum | Desired |
|---|---|---|
| DR | 22 dB (160:1) | ≥ 25.4 dB (350:1) |
| Lmax[a] | 103 cd/m$^2$ (30 fL) | ≥ 120 cd/m$^2$ (35 fL) |
| Luminance nonuniformity[b] | ≤ 20% | ≤ 10% |
| Viewing angle threshold[c] | – 3 dB @ ± 12 deg | – 3 dB @ ± 30 deg |
| Halation | ≤ 3.5% | ≤ 2% |
| Bit depth[d] | 8 bit | 10 bit |
| Color temperature | D65 ≥ T ≤ D93 | Same |
| Color uniformity | Δ u' v' ≤ 0.01 | Same |

[a] Monochrome requirement applies for non-CRT displays. The American College of Radiology (ACR) recommends ≥ 50 fL.
[b] A value of ≤ 15% has been recommended for radiologist displays.
[c] For AMLCD displays.
[d] The ACR recommends 10 bits for radiography, but 8 bits is considered adequate for CT and MRI images.

Minimum performance levels are based on results of information extraction studies as well as a review of the marketplace. Thus, a display with no perceptible luminance nonuniformity would be preferable, but such displays do not currently exist. The minimum value listed is thus one that is achievable in the current market. Minimum values may also represent performance thresholds where studies have shown a significant decrease in information extraction performance at values below the minimum. The 22-dB DR minimum, for example, represents a performance threshold. The desired values are sometimes based on results of previous studies but are usually based on theoretical grounds or engineering considerations. As an example, if a perceptual linearization LUT is applied to an 8-bit system, some degree of quantization results and the number of unique luminance levels is reduced by as much as 20%. With a 10-bit digital-to-analog converter (DAC), the number of levels is still greater than the number of perceptually just-noticeable contrast differences, even after a LUT is applied. However, the performance benefit of a 10-bit DAC has not been conclusively demonstrated in the literature to date.

In some cases, minimum or desired values vary depending on the application. For example, the medical literature suggests a need for higher luminance and DR than the aerial imagery literature. In the discussion that follows, the rationale for each requirement is provided.

## 7.2 Measurement Definition

Dynamic range is the ratio of Lmax to Lmin. It is expressed as a contrast ratio—e.g., 300:1—or in decibels (dB). Dynamic range in dB is 10 times the $\log_{10}$ of the

# CONTRAST PERFORMANCE REQUIREMENTS

**Table 7.2** Luminance requirements–Monochrome.

| Measure | Minimum | Desired |
|---|---|---|
| DR | $\geq$ 25.4 dB (350:1) | $\geq$ 25.4 dB (350:1) |
| Lmax[a] | 120 cd/m$^2$ (35 fL) | $\geq$ 120 cd/m$^2$ (35 fL) |
| Luminance nonuniformity[b] | $\leq$ 20% | $\leq$ 10% |
| Viewing angle threshold[c] | – 3 dB @ ± 12 deg | – 3 dB @ ± 30 deg |
| Halation | $\leq$ 3.5% | $\leq$ 2% |
| Bit depth[d] | 8 bit$^2$ | 10 bit |

[a] The American College of Radiology (ACR) recommends $\geq$ 171 cd/m$^2$ (50 fL).
[b] A value of $\leq$ 15% has been recommended for radiologist displays.
[c] For AMLCD displays.
[d] The ACR recommends 10 bits for radiography, but 8 bits is considered adequate for CT and MRI images.

**Table 7.3** Luminance requirements–Stereo.

| Measure | Minimum | Desired |
|---|---|---|
| DR—monochrome | 22.7 dB | $\geq$ 25.4 dB |
| DR—color | 17.7 dB | $\geq$ 25.4 dB |
| Lmax—monochrome | 103 cd/m$^2$ (30 fL) | $\geq$ 120 cd/m$^2$ (35 fL) |
| Lmax—color | 21 cd/m$^2$ (6 fL) | $\geq$ 120 cd/m$^2$ (35 fL) |

contrast ratio. Maximum luminance is the highest luminance value that can be achieved by the monitor, which is usually in the center for a CRT. Luminance nonuniformity is the maximum percentage difference in Lmax as a function of screen position. With DR and Lmax specified, Lmin is redundant.

The viewing angle threshold is the allowable change in DR over a prescribed viewing angle. Halation is the percentage luminance increase in a small black area (0 CL) when surrounded by the maximum commanded luminance level ($CL_{max}$). Color temperature is the temperature of a blackbody radiator with the same color coordinates for white as the display at maximum luminance. Bit depth is the $\log_2$ of the number of CLs. Color uniformity is a measure of the degree to which color coordinates are constant across the face of a display.

## 7.3 Requirement Rationale

In terms of documented evidence, the DR and Lmax requirements are the most important luminance requirements. The combination of DR and Lmax defines the number of potentially available luminance JNDs. Using the NEMA/DICOM model,[8] Fig. 7.1 shows the effect of DR and Lmax on the number of luminance JNDs. It is apparent that increasing Lmax is more beneficial than increasing DR by

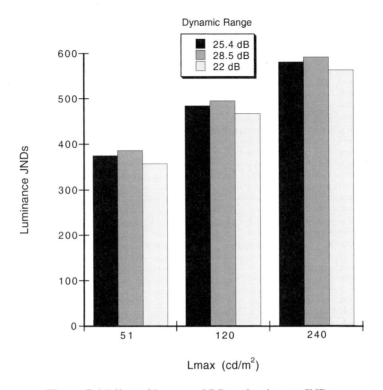

**Figure 7.1** Effect of Lmax and DR on luminance JNDs.

decreasing Lmin. However, at least three factors modify the data shown in Fig. 7.1. First, CRTs have upper limits on Lmax. For color CRTs, that limit is currently on the order of 120 cd/m$^2$ (35 fL). For monochrome CRTs, as luminance increases the pixels grow in size (called spot growth), thus diminishing resolution. There is also a visual comfort limit on Lmax. An AMLCD monitor with an Lmax of 774 cd/m$^2$ (226 fL) was judged as too bright by some observers when viewed in a 194-lx illuminance setting.[9] If the screen were filled with imagery, the high brightness would not be apparent, thus suggesting that the high brightness would be acceptable if the screen were always filled with imagery. This would require software that would command areas of the screen not occupied with image windows to a low luminance level. A second factor modifying the data in Fig. 7.1 is an undefined relationship with bit rate. An 8-bit display has only 255 commandable steps, so any potential JNDs beyond that number will not be physically realized. Third, maintaining DR while reducing Lmax is often not possible. Any room light adds to Lmin, so very low values of Lmin cannot be achieved.

Figure 7.2 shows results of a study where the Lmax and DR of a color monitor (100 ppi at 2x magnification) were varied. Multispectral NIIRS ratings (A) and color Briggs ratings (B) were obtained.[10] Decreasing Lmax had a somewhat greater effect than decreasing DR, but a significant NIIRS loss occurred between 24 and 18 dB. In this and all succeeding graphs, data values connected by a line labeled NS do not differ by a statistically significant amount.

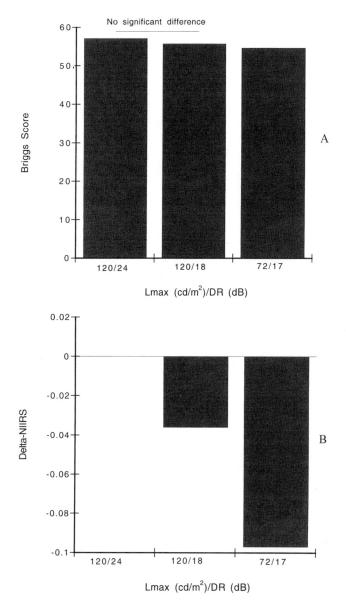

**Figure 7.2** Effect of Lmax and DR on Briggs and NIIRS ratings.

Results of a second study using a monochrome monitor are shown in Fig. 7.3.[11] Increasing Lmax and DR, or DR alone, had no significant effect on Briggs scores or delta-NIIRS. Note that according to the Barten model described in Sec. 6.4.1, the number of discriminable gray levels for the 0.34- to 120-cd/m$^2$ (0.1 to 35 fL) calibration exceeds 256 (8 bits).

At least three studies in the medical literature have addressed the DR/Lmax issue. The standard for the medical community has been the high-intensity light box used to display radiographic images. The intensity of a light box can be on the

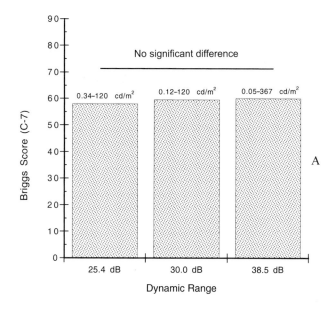

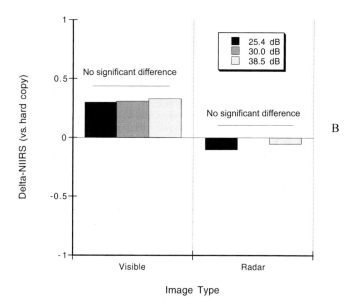

**Figure 7.3** Effect of DR alone on Briggs and NIIRS ratings.

order of 2055 to 3426 cd/m$^2$ (600 to 1000 fL). The density of the film to be viewed reduces this to a DR on the order of 25 to 28 dB[12] or a luminance range of perhaps 3.4 to 2400 cd/m$^2$ (1 to 700 fL), assuming a film density range of 0.2 to 3.0. In a comparison of 65-fL and 100-fL (Lmax) monitors (223–343 cd/m$^2$),[13] the detection of pulmonary nodules by experienced radiologists on 80 chest images did not differ significantly. Performance, as determined by ROC analysis, was greater on the 100-fL monitor. In a study in which simulated masses were embedded in mammograms, Lmax (using neutral density filters) varied over a range of 10 to 600 fL (34.3 to 2055 cd/m$^2$).[14] However, differences in Lmax did not produce statistically significant performance differences. In a third study, observers searched for lesions on mammograms displayed on two monitors, one with an Lmax of 80 fL (274 cd/m$^2$) and the other with an Lmax of 140 fL (480 cd/m$^2$).[15] Although the detection/false alarm rates did not differ significantly between the two monitors, less time was required on the 140-fL monitor. One could thus argue that lower luminance might contribute to a fatigue effect because of the added viewing time required. The ACR Standard for Teleradiology recommends a value of $\geq$ 50 fL (171 cd/m$^2$) for Lmax.[16] Note, however, that without a specification on DR, the potential number of gray levels is undefined.

Results of the medical studies, although not conclusively supporting high luminance levels, certainly do not argue otherwise. Experience with hard copy indicates the importance of luminance levels. We thus conclude that for medical applications, a DR of 25 dB or greater is desirable with an Lmax that is as high as possible without degrading resolution due to spot growth. Similarly, with stereo, one would like to obtain the same output through the viewing system as with unaided viewing. This is not possible with current stereo viewing devices, hence the lower values.

The effect of limiting DR on aerial imagery is illustrated in Fig. 7.4. An image with full DR is shown on the left with a histogram of the image displaying the

**Figure 7.4** Effect of limiting DR.

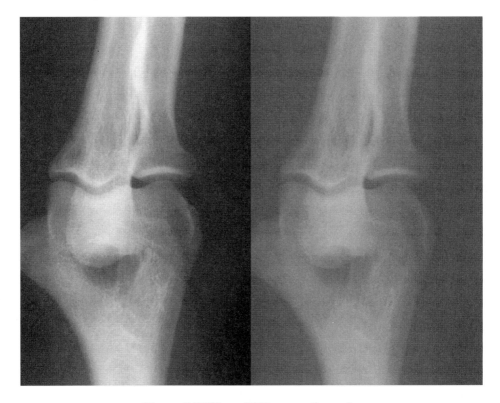

**Figure 7.5** Effect of DR on a radiograph.

number of pixels at each CL. On the right, the same image with half the DR and a reduced number of CLs is shown, resulting in an image with a "flatter" appearance.

Figure 7.5 shows the effect of limiting DR on a radiograph. The image on the left has a 24-dB DR; the DR of the image on the right is 15 dB. This is equivalent to a monitor with a DR of 1.7 to 61.7 cd/m$^2$ (0.5 to 18 fL). It is not uncommon for aging and uncalibrated (dark cutoff) color monitors to show such a restricted DR.

Totally uniform luminance as a function of position, although desirable, is not achievable with electronic displays. It is also not possible with light boxes or light tables used for hard-copy viewing. In theory, one could compensate in software for luminance nonuniformity by developing LUTs based on screen position. Some degree of compensation is also possible in hardware. A display using a digital controller with almost perfect luminance uniformity has reportedly been produced.[7] An achievable figure with current monitors is 20%, while 10% is a goal. A value of ≤ 15% has been recommended for radiologists' monitors and ≤ 20% for clinicians.[17]

Figure 7.6 shows two vehicles labeled A and B. The rooftops of the two vehicles differ in luminance by about 20%, but the difference in tone is barely discernible. Therefore, the real luminance uniformity issue would seem to be one of avoiding significant changes in DR. Measurements made on a low-cost color monitor showed a range of 4 dB in DR. All DR values were greater than 28 dB.

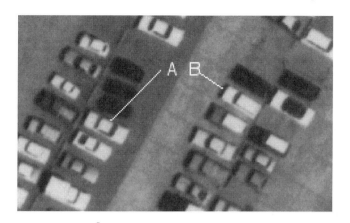

**Figure 7.6** Effect of 20% luminance variation.

The viewing angle threshold specification has two components—one of luminance and one of size. The luminance component is actually a DR specification and places a limit on the loss of DR. The ±12-deg specification describes an area on the display 8 in. in diameter at an 18-in. viewing distance; the ±30-deg specification describes a circle that encompasses almost all of the full screen on a typical 4:3 aspect ratio, 21-in. CRT display.

The halation specification is again based on current technology and is somewhat redundant with the Cm value that will be defined in the following chapter. Halation also reduces DR in a manner not captured in the standard method of defining DR. Halation has been measured in a variety of ways. The NIDL[3,6] measures the ratio of the luminance of a small black square in a white surround to the difference in luminance levels between a small black area and a small white area. The VESA[1] measures halation as the increase in black square luminance as a percentage of Lmax. In either case, the luminance of a black square or area increases as the surround becomes brighter and/or larger. The luminance of the black area also varies with its size. It is thus difficult to interpret a single value since it is subject to so many measurement conditions—variations of 100% as a function of measurement definition and condition are common. Other than a desire to limit halation in order to avoid contrast reduction, there is no absolute basis for the specification. Minor violations should not be cause to reject a display.

The minimum requirement for bit depth represents what is currently available on PCs. The 10-bit desired value is based on theory and is recommended by the ACR for radiography.[18] Both 10- and 12-bit controllers are available in the market. The 8-bit value is considered adequate by the ACR for small matrix images such as computed tomography (CT) and magnetic resonance imaging (MRI). In many cases, it is difficult to see the difference in bit depth by the appearance of an image. For example, Fig. 7.7 shows an image at three bit depths, but differences among the three are difficult to discern. Histograms of the three images are shown in Fig. 7.8. Differences in bit depth become more apparent in large flat fields, where a phenomenon called contouring can occur. Contouring appears as gray-level steps

**Figure 7.7** An image at three bit depths.

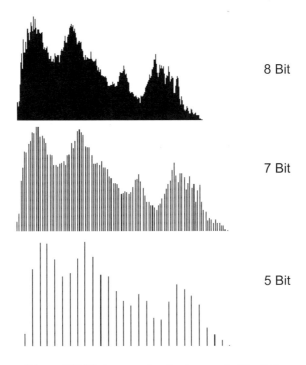

**Figure 7.8** Histograms for the images in Fig. 7.7.

rather than a continuous transition. Figure 7.9 shows an example. Both images have been enlarged by a factor of 4. It should be noted that any loss of gray levels is potentially harmful, but the attendant loss of contrast may not be obvious.

From Fig. 7.1, it must be recognized that the number of potentially available luminance JNDs exceeds the number of CLs in even an 8-bit display. Further, when a perceptual linearization LUT is applied, levels are lost to quantization. Assuming some 600 JNDs (from Fig. 7.1), a 10-bit display should be capable of maintaining the JNDs even with quantization loss. Note that there are 12-bit controllers currently on the market that could potentially be beneficial in maintaining the integrity of 11-bit and higher original images.

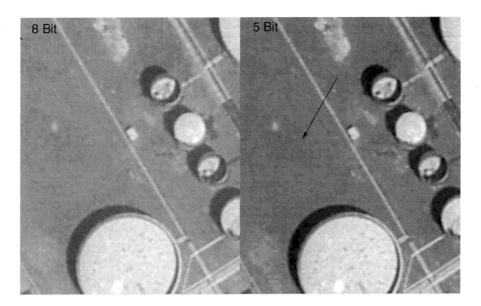

**Figure 7.9** Illustration of contouring.

Recommended color temperature is based on a study in which Briggs and NIIRS ratings were made on color Briggs targets and multispectral imagery, respectively.[10] No difference was seen between the two values in terms of performance.

Luminance is generally higher at the higher color temperature. Thus, it is often useful to set the color temperature at D93 (9300 K) in order to obtain the highest possible Lmax value. The color uniformity specification is based on what is routinely achieved with color CRT displays. Differences of this magnitude are not visually apparent. Note from Chapter 6 that larger differences will not be discriminable in real imagery.

Under one particular set of circumstances, it may be desirable to fix the color temperature of a display at a particular value. This is a situation where multiple displays are used in close proximity to one another. Monitors set to different color temperatures will have different color casts (red for D65, blue for D93) and this difference may be annoying or create the impression that one monitor is "better" than another. This phenomenon has also been observed with monochrome monitors where phosphor differences produce color temperature differences. Although the two monochrome monitors did not differ in terms of performance, observers expressed strong (but inconsistent) preferences for one over the other.[19]

The Lmax and DR requirements (minimum levels) for stereo viewing are based on what is reasonably achievable with current technology. Stereo viewing devices extract a heavy penalty in luminance, and the impact of that loss has not yet been assessed. Desired performance is assumed to be the same as that for monoscopic viewing.

Performance requirements have not been defined for several of the measures discussed in Chapter 4. They include gamma, the I/O function, reflectance, luminance stability, and gamut. As noted previously, gamma is typically indeterminate since the I/O function in log/log space is nonlinear. Further, gamma is somewhat redundant with Lmax and DR. In Chapter 10, a procedure for developing a LUT to modify the I/O function to a desired function is defined, and thus no requirement is needed here. In the proper operating environment (low light level), reflectance is generally unimportant since there is almost no light to be reflected. Further, a low-reflectance screen reduces transmission and thus Lmax and DR. For that reason, no requirement is provided. If the user's intention is to operate the display in a well-lit environment, it will be useful to measure the DR in that environment to assess the loss due to the lighting. Luminance stability covers a variety of measures. Instability will be reflected in Cm measurements. Finally, gamut is largely a function of the material used to produce color on the display (e.g., phosphor type) as well as the DR or Lmax of the display. Although a large gamut is theoretically beneficial, there is no basis for setting a gamut requirement separate from an Lmax requirement.

## 7.4 Instrument Measurement

In this section, a set of procedures for measuring each of the contrast performance parameters is outlined. With noted exceptions, all of the measurements should be made at the system level. The display should be driven by the workstation with which it will be used. This procedure differs from the normal procedure of using a signal generator to test a monitor, because here, the performance of the monitor as a system component is of interest. Both hardware and software external to the monitor may affect its performance.

The monitor should be allowed to warm up until Lmax stabilizes at a constant value. The variation in luminance before a monitor has warmed up is considered a source of noise and is treated in Chapter 9. Warm-up time can be as long as one hour.

Measurements should be made in a darkened environment or with the monitor completely shielded from any room light or reflections (e.g., with a heavy black cloth). The monitor/workstation configuration should be held constant during the measurement process. CRT's can be sensitive to magnetic fields that are affected by other equipment. Rotation of a monitor, for example, may change color purity and, in extreme cases, luminance readings.

Readings are also sensitive to characteristics of the device used to make measurements. Measurement devices must be calibrated to a reference source, typically on at least a yearly basis. Lens flare is a problem with certain types of measuring devices. The reader is referred to Ref. [1] for a more detailed treatment of issues and procedures.

The procedures that follow were drawn from Refs. [1]–[8]. In some cases they have been modified, particularly with an aim toward simplification. Major differences are noted. With the exception of bit depth, all of the required measurement targets are included on the attached CD.

## 7.4.1 Initial setup

*Objective:* Set up the monitor in preparation for all other measurements.

*Equipment:* A photometer that can read from 0.03 cd/m² (0.01 fL) to 686 cd/m² (200 fL) with accuracy of ±5%.

*Procedure:* Set the display format to the desired addressability (normally the highest addressability). Set brightness and contrast controls at manufacturer's recommendation. If no recommendation is provided, define a target with a box centered in a full-screen display. The box should have a dimension of 10% of the viewable screen area (~4 × 4 in. for a 21-in. CRT). Set the box at CL 1 and the background at level 0. Measure the box and background to ensure that there is a measurable luminance difference. Adjust the display brightness control until the difference is just measurably detectable, or, alternatively, adjust the background to a setting of 0.3 cd/m² (0.1 fL) and ensure that the box has a measurably higher luminance value. Adjust contrast to the maximum setting. Command the full screen to the maximum CL (CLmax). Measure the luminance every minute for the first 30 minutes or until luminance stabilizes to ± 10%. If luminance is not stable within 30 minutes, extend the measurement interval to every 5 minutes for the next 30 minutes and every 10 minutes thereafter.

*Analysis:* Plot CLmax luminance readings and determine when readings are constant to within ±1%.

## 7.4.2 Dynamic range

*Objective:* Measure luminance as a function of CL at CLmax and Clmin, where CLmax and CLmin are the maximum and minimum display CLs (0 and 255 for an 8-bit display).

*Equipment:* A photometer that can read from 0.03 cd/m² (0.01 fL) to 686 cd/m² (200 fL) with accuracy of ±5%.

*Procedure:* Set the color temperature at the desired value. Lmax will be maximized when the temperature is set at D93. Maximizing the gain of each color gun will also maximize Lmax. Command the full display to 0. Measure luminance in the center of the screen. Command the full screen to CLmax. Measure luminance in the center of the screen. For some systems, it may not be possible to command the full screen because of the presence of menus. In this case, command the maximum possible area and ensure that the portion not commanded to the desired value does not interfere with the measurement.

*Analysis:* Define DR by Lmax/Lmin:1 or $10\log_{10}(Lmax/Lmin)$.

### 7.4.3 Lmax

*Objective:* Measure the maximum display luminance.
*Equipment:* A photometer that can read from 0.03 cd/m² (0.01 fL) to 686 cd/m² (200fL) with accuracy of ±5%.
*Procedure:* Using the procedures defined in DR, record the Lmax value.
*Analysis:* Lmax is the recorded maximum luminance value.

### 7.4.4 Input/output function

*Objective:* Measure the relationship between input CL and output luminance. This function is required for subsequent measurements. Note that it may be desirable to set Lmax to a defined value (e.g., 200 or 170 cd/m²) as opposed to the maximum possible value. The value used should be that at which the display will be operated.
*Equipment:* A photometer that can read from 0.03 cd/m² (0.01 fL) to 686 cd/m² (200 fL) with accuracy of ±5%.
*Procedure:* Define a test target that consists of a box within a full-screen background. The box should take on the successive values shown in Table 7.4. For an 8-bit display, use the values of 0 to 255. A 10-bit display uses values up to 1024, and a 12-bit display uses values up to 4096. The box should have a height and width that is 10% of the full screen area (about 4 × 4 in. for a 21-in. display). For photometers that use a hood or puck, the box should be at least 10% larger than the hood or puck. The background should be commanded to the values shown in Table 7.4. Measure the output luminance for each of the boxes.
*Analysis:* Plot CL vs. luminance as a log/log function. Define the luminance values that are 75%, 50%, and 25% of the maximum (at CLmax). Read off the CLs corresponding to those luminance levels. Alternatively, define luminance as a function of CL using a third-order polynomial such that

$$L = a = bCL + cCL^2 + dCL^3. \tag{7.1}$$

Save this function for subsequent use in monitor calibration.

### 7.4.5 Luminance uniformity

*Objective:* Measure the variation in output luminance as a function of screen position and intensity.
*Equipment:* A photometer that can read from 0.03 cd/m² (0.01 fL) to 686 cd/m² (200 fL) with accuracy of ±5%.

*Procedure:* In order to measure luminance uniformity, it is necessary to know the CLs corresponding to luminance values of 25%, 50%, and 75% of Lmax. These were defined in the previously described I/O function measurement procedure. Set the CL to CLmax and measure luminance at the nine points defined in Fig. 7.10. Set the CL to the value corresponding to 75% of Lmax and repeat the measurements. Repeat again for the 50% and 25% values.

*Analysis:* At each of the four CL values, define the maximum and minimum measured luminance values. Luminance uniformity at a given percent of Lmax is equal to

$$\text{Luminance nonuniformity} = 100 \times \left(\frac{Lmax - Lmin}{Lmax}\right), \quad (7.2)$$

where *Lmax* and *Lmin* are the values measured at each of the four CLs.

Table 7.4 Box and background CLs.

| Box | Background |
|---|---|
| 0 | 38 |
| 1 | 39 |
| 2 | 39 |
| 3 | 39 |
| 4 | 40 |
| 6 | 40 |
| 8 | 41 |
| 11 | 42 |
| 16 | 44 |
| 23 | 46 |
| 32 | 49 |
| 45 | 54 |
| 64 | 61 |
| 91 | 70 |
| 128 | 83 |
| 181 | 102 |
| 255 | 128 |
| 361 | 165 |
| 512 | 217 |
| 723 | 291 |
| 1024 | 397 |
| 1447 | 545 |
| 2048 | 755 |
| 2895 | 1052 |
| 4096 | 1472 |

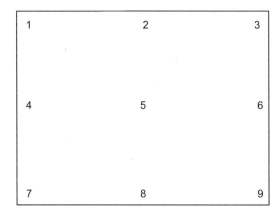

**Figure 7.10** Measurement locations. Location 5 is centered on the display; all other locations are 1 in. from the edge of the viewable area (or a distance equal to 10% of the screen height or width, or both for corner measurements).

## 7.4.6 Viewing angle threshold

*Objective*: Measure the decrease in DR as a function of viewing angle.

*Equipment*: A spot photometer that can read from 0.03 cd/m$^2$ (0.01 fL) to 686 cd/m$^2$ (200 fL) with accuracy of ±5% and an angular measurement device (goniometric positioning device).

*Procedure*: Command the full screen to Lmax. Measure luminance at center and at 12 deg from center (3.8 in.) and 1 in. from corner (or 30 deg if less than 1 in. from the corner) as shown in Fig. 7.11. Repeat for each of the four quadrants of the display. Command the full screen to Lmin and make luminance measurements at the same points.

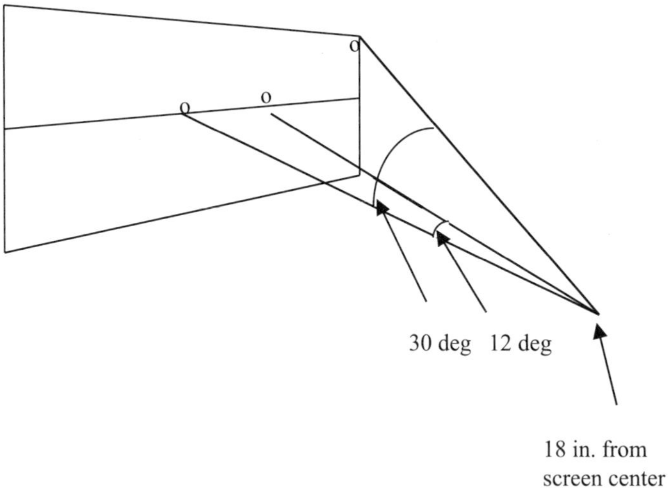

30 deg   12 deg

18 in. from screen center

**Figure 7.11** Viewing angle measurements.

*Analysis:* Compute the DR at each position. Define the difference between DRmax and DRmin. The difference should be less than 3 dB [defined as DRmax(dB)–DRmin(dB)].

## 7.4.7 Halation

*Objective:* Measure contrast degradation due to halation.
*Equipment:* A photometer that can read from 0.03 cd/m² (0.01 fL) to 686 cd/m² (200 fL) with accuracy of ±5% and 4-in. black cardboard mask (if needed based on photometer field of view).
*Procedure:* Define a target consisting of a box centered in a full-field display. The box should have a dimension of 5% of the full screen area (3 in. for a nominal 21-in. CRT). If the box is less than twice the photometer field of view, cut a hole in the mask that is half the dimension of a 5% box. Place the mask over the box (centered) for measurements. The box should be commanded to CLmin and the surround to CLmax. There should be no ambient light on the screen. Measure the luminance of the box. An alternative procedure calls for increasing the box size incrementally to 100% of the screen size and measuring the luminance for each box.[1] Lmax is defined as the highest value read (which is normally that of the smallest box).
*Analysis:* Define halation as

$$\text{Halation}(\%) = 100\left(\frac{L_{black}}{L_{max} - L_{min}}\right), \quad (7.3)$$

where $L_{black}$ is the luminance of the black box, $L_{max}$ is the luminance of the full screen at CLmax, and $L_{min}$ is the luminance of the full screen at CLmin.

## 7.4.8 Bit depth

*Objective:* Define the number of bits of data that can be displayed.
*Equipment:* A photometer that can read from 0.03 cd/m² (0.01 fL) to 686 cd/m² (200 fL) with accuracy of ±5%.
*Procedure:* Define targets consisting of a box centered in a full-screen display. The box should be 10% of the full screen size area or 10% larger than the photometer hood or puck. The box should be commanded to all possible values, e.g., 0–1023 for a 10-bit display. The box background should be commanded to

$$CL_{Bkgd} = 0.5[(0.7 \times CL_{Box}) + 0.3n], \quad (7.4)$$

where *n* is the number of commandable levels. Measure the luminance of each box at the center.

*Analysis:* Determine the number of unique luminance levels. Bit depth is $\log_2$ times the number of unique luminance levels.

## 7.4.9 Color temperature

*Objective:* Verify the color temperature of the monitor.

*Equipment:* A colorimeter or spectroradiometer with a spectral range of 400 to 700. The spectroradiometer's accuracy should be ±1 nm and CIE *x*, *y* coordinate measurement accuracy should be within 0.003 units. Colorimeter accuracy for the CIE *x*, *y* coordinate values should be within 0.006 units.

*Procedure:* Command the full screen to Lmax. Measure CIE *x*, *y* coordinates at the screen center.

*Analysis:* Coordinates should fall on a line defined by a third-order polynomial fit to the 5000 to 10,000 K *x*, *y* coordinates and falling on or between the D65 and D93 coordinates with a tolerance of ± 0.01Δ *u'v'*. The *x*, *y* coordinates for a D illuminant are defined by two equations for the *x* coordinate (as a function of the color temperature) and one for the *y* coordinate (as a function of the *x* coordinate).[20] For temperatures between 4000 K and 7000 K,

$$x = -4.6070\frac{10^9}{T^3} + 2.9678\frac{10^6}{T^2} + 0.0991\frac{10^3}{T} + 0.244063; \quad (7.5)$$

for temperatures between 7000 K and 25,000 K,

$$x = -2.0064\frac{10^9}{T^3} + 1.9018\frac{10^6}{T^2} + 0.24748\frac{10^3}{T} + 0.237040. \quad (7.6)$$

The value for *y* is computed from *x* as

$$y = -3.000x^2 + 2.870x - 0.275. \quad (7.7)$$

For a range of *T* between 5000 K and 10,000 K, define the *x*, *y* coordinates. Fit a third-order polynomial to the resultant data points.

### 7.4.10 Color uniformity

*Objective:* Define color variations as a function of position and luminance level.

*Equipment:* A colorimeter or spectroradiometer with a spectral range of 400 to 700. Spectroradiometer accuracy should be ±1 nm and CIE $x$, $y$ coordinate measurement accuracy should be within 0.003 units. Colorimeter accuracy for CIE $x$, $y$ coordinate values should be within 0.006 units.

*Procedure:* Measure color coordinates ($u'v'$) at positions and luminance levels as defined by luminance uniformity measurement.

*Analysis:* Define $\Delta u'v'$ for each measurement relative to the screen center at Lmax. Define the maximum $\Delta u'v'$.

## 7.5 Measurement Alternatives

Some organizations and individuals may not have the resources or desire to adopt a fully instrumented approach to display measurement. In this section, we explore alternatives to luminance and color measurement.

The popular press has generally stopped reviewing and reporting on specific displays. The rapid changes in models and the expense of conducting detailed measurements have perhaps contributed to this situation. In the area of luminance data, vendors may at most define Lmax and contrast ratio; and for AMLCDs, the viewing angle. The conditions under which the measurements are made are generally undefined and, as noted previously, the viewing angle value may be quite different than that obtained using the procedures described here. The NIDL does currently make detailed measurements of monitors available to government users at http://www.nidl.org/tech_mstr.htm.

Most users should be able to afford a minimum set of luminance and illuminance measurement equipment. A puck-type luminance measurement device can be purchased for under $1000 (U.S.) and an incident illumination light meter for a few hundred dollars.

A light-measuring device such as a camera with an exposure meter, or an exposure meter itself, can be used to obtain an approximate indication of DR. The full screen is commanded to CLmin and CLmax with no ambient light. The exposure value (EV), or the aperture and the exposure, are measured at both settings. Each successive EV change is equivalent to a doubling or halving of light energy. The exposure value difference is computed from Table 7.5. An EV difference of 8–9 provides a reasonable DR.

Lmax cannot be measured without a photometer. A piece of white paper 18 in. from a 60-W soft white bulb measures ~ 62 cd/m$^2$, and at 12 in., ~ 127 cd/m$^2$. The desired Lmax should be considerably brighter than 62 cd/m$^2$. Significant luminance nonuniformity can be detected with the target shown in Fig. 7.12. Six step wedges of seven steps each (255–225 in increments of five CLs) are arrayed around a full-screen image. Viewing angle limits can be assessed with the target in

Fig. 7.13. The target should be moved around the display. Any appreciable change in contrast indicates the viewing angle limit may have been reached. A viewing angle target using color has also been developed.[21]

**Table 7.5** Exposure value table.

| Speed/f# | 1 | 2 | 3 | 4 | 6 | 8 | 11 | 16 | 22 | 32 | 45 | 64 |
|---|---|---|---|---|---|---|---|---|---|---|---|---|
| 8 | −3 | −2 | −1 | 0 | 1 | 2 | 3 | 4 | 5 | 6 | 7 | 8 |
| 4 | −2 | −1 | 0 | 1 | 2 | 3 | 4 | 5 | 6 | 7 | 8 | 9 |
| 2 | −1 | 0 | 1 | 2 | 3 | 4 | 5 | 6 | 7 | 8 | 9 | 10 |
| 1 | 0 | 1 | 2 | 3 | 4 | 5 | 6 | 7 | 8 | 9 | 10 | 11 |
| 1/2 | 1 | 2 | 3 | 4 | 5 | 6 | 7 | 8 | 9 | 10 | 11 | 12 |
| 1/4 | 2 | 3 | 4 | 5 | 6 | 7 | 8 | 9 | 10 | 11 | 12 | 13 |
| 1/8 | 3 | 4 | 5 | 6 | 7 | 8 | 9 | 10 | 11 | 12 | 13 | 14 |
| 1/15 | 4 | 5 | 6 | 7 | 8 | 9 | 10 | 11 | 12 | 13 | 14 | 15 |
| 1/30 | 5 | 6 | 7 | 8 | 9 | 10 | 11 | 12 | 13 | 14 | 15 | 16 |
| 1/60 | 6 | 7 | 8 | 9 | 10 | 11 | 12 | 13 | 14 | 15 | 16 | 17 |
| 1/125 | 7 | 8 | 9 | 10 | 11 | 12 | 13 | 14 | 15 | 16 | 17 | 18 |
| 1/250 | 8 | 9 | 10 | 11 | 12 | 13 | 14 | 15 | 16 | 17 | 18 | 19 |
| 1/500 | 9 | 10 | 11 | 12 | 13 | 14 | 15 | 16 | 17 | 18 | 19 | 20 |

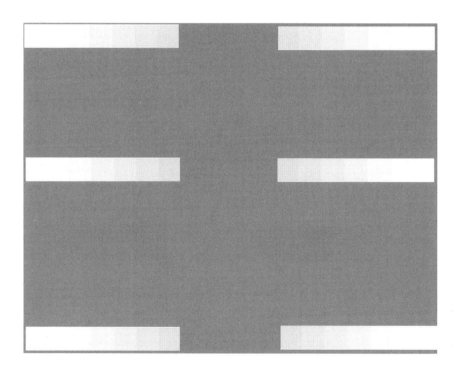

**Figure 7.12** Luminance nonuniformity target.

# CONTRAST PERFORMANCE REQUIREMENTS

**Figure 7.13** Viewing angle target.

**Figure 7.14** Color uniformity target.

Color nonuniformity may be detected with the target shown in Fig. 7.14. Any appreciable change in color between or within the four squares may be cause for concern depending on the application.

Halation and bit depth cannot be measured without a photometer. The required accuracy precludes estimation through mere observation. This is also the case for bit depth and the I/O function. Color temperature requires a colorimeter for verification. The appendix provides two targets that can be used to visually assess monitor color performance at a gross level.

## 7.6 Summary

Lmax and DR represent the two most important luminance requirements for a monitor. For color CRT displays, Lmax is limited to about 120 cd/m$^2$ (35 fL). For monochrome CRTs, higher Lmax values can be achieved, but sometimes at the expense of spot growth that degrades resolution. Dynamic range should be maintained above 22 dB; values of greater than 25 dB have shown relatively little effect on performance. Ideally, luminance should be uniform across the display, but this goal can never be achieved. The performance requirements provided in this chapter represent realizable goals. With AMLCDs, viewing angle is an issue; contrast and color vary with off-axis viewing. Bit depth defines the number of uniquely commandable gray levels or colors, but the number actually achieved is generally reduced with the application of a perceptual linearization LUT (yet overall performance is improved). Halation reduces both contrast and achievable Cm; the requirement provided ($\leq 3.5\%$) is a technology-based value. Color temperature appears to have no impact on performance over the range of D65 to D93. Higher Lmax values can be achieved, however, at the D93 temperature. Current displays show good color uniformity, so maintaining color uniformity is usually an issue only when multiple displays are used in close proximity.

Ensuring compliance with the luminance requirements provided requires a photometer and a colorimeter or spectroradiometer. A photometer can be purchased for under $1000 (U.S.) and is strongly recommended for organizations with multiple workstations. Measurements must be made in controlled conditions after appropriate monitor setup. In the absence of the required instrumentation, only general approximations can be made. With the exception of data provided by the NIDL, vendor data are incomplete and may be obtained under conditions and procedures different from those recommended here.

## References

[1] Video Electronic Standards Association, *Flat Panel Display Measurements Standard*, Version 2.0, VESA, Milpitas, CA (2001).
[2] National Information Display Laboratory, *Display Monitor Measurement Methods under discussion by EIA Committee JT-20, Part 1, Monochrome CRT Monitor Performance,* Version 2.0, Princeton, NJ (1995).

[3] National Information Display Laboratory, *Display Monitor Measurement Methods under discussion by EIA Committee JT-20, Part 2, Color CRT Monitor Performance,* Version 2.0, Princeton, NJ (1995).

[4] National Information Display Laboratory, *Test Procedures for Evaluation of CRT Display Monitors,* Version 2.0, Princeton, NJ (1991).

[5] J. Leachtenauer, *SC Measurement Procedures,* NIMA, Reston, VA (1998).

[6] National Information Display Laboratory, *Request for Evaluation Monitors for the National Imagery and Mapping Agency (NIMA) Integrated Exploitation Capability (IEC,)* Princeton, NJ (1999).

[7] H. Roehrig, "The monochrome cathode ray tube display and its performance," in Y. Kim and S. Horii (eds.), *Handbook of Medical Imaging, Vol. 1, Physics and Psychophysics,* Bellingham, WA, pp. 157–220 (2000).

[8] National Electrical Manufacturers Association, *Digital Imaging and Communications in Medicine (DICOM), Part 14: Grayscale Standard Display Function,* Rosslyn, VA (2001).

[9] J. Leachtenauer, A. Biache, and G. Garney, *Comparison of AMLCD and CRT Monitors for Imagery Display,* Technology Assessment Office, National Imagery and Mapping Agency, Reston, VA (1998).

[10] J. Leachtenauer and N. Salvaggio, " Color monitor calibration for display of multispectral imagery," Society for Information Display, International Symposium, *Digest of Technical Papers,* Boston, MA, May 13–15, 1997, pp. 1037–1040.

[11] J. Leachtenauer, A. Biache, and G. Garney "Effects of ambient lighting and monitor calibration on softcopy image interpretability," *Final Program and Proceedings, PICS Conference,* Society for Imaging Science and Technology, Savannah, GA, April 25–28, 1999, pp. 179–183.

[12] E. Muka, H. Blume, and S. Daly, "Display of medical images on CRT softcopy displays: a tutorial," *Proc. SPIE,* Vol. 2431, pp. 341–358 (1995).

[13] K.-S. Song, J. S. Lee, H. Y. Kim, and T.-H. Lim, "Effect of monitor luminance on the detection of solitary pulmonary nodule: ROC analysis," *Proc. SPIE,* Vol. 3663, pp. 212–216 (1999).

[14] B. M. Hemminger, A. Dillon, and R. E. Johnston, "Evaluation of the effect of display luminance on the feature detection rates of masses in mammograms," *Proc. SPIE,* Vol. 3036, pp. 96–106 (1997).

[15] E. A. Krupinski, H. Roehrig, and T. Furukawa, "Influence of film and monitor display luminance on observer performance and visual search," *Acad. Radiol.,* Vol. 6, pp. 411–418 (1999).

[16] American College of Radiology, *ACR Standard for Teleradiology,* Res. 35, (1999).

[17] J. F. Copeland et al., "Practical quality control standards for digital display monitors," *Proc. SPIE,* Vol. 3976, pp. 315–322 (2000).

[18] American College of Radiology, *ACR Standard for Digital Image Data Management,* Res. 6 (2002).

[19] J. Leachtenauer, G. Garney, and A. Biache, "Contrast modulation—how much is enough?" *Final Program and Proceedings, PICS Conference,* Soci-

ety for Imaging Science and Technology, Portland, OR, March 26–29, 2000, pp. 130–134.

[20] G. Wyszecki and W. S. Stiles, *Color Science*, Second Edition, John Wiley & Sons, Inc., New York (1982).

[21] M. H. Brill, "Color reversal at a glance," *Information Display*, Vol. 6, pp. 36–37 (2000).

# Chapter 8
# Size and Resolution Performance Requirements

The ability to see fine detail on a monitor is a function of various size-related parameters. These parameters ultimately define the Cm performance of the display—the ability to see small, low-contrast detail. This chapter provides a set of size/resolution specifications and associated measurement procedures, beginning with a listing of the requirements and their definitions followed by a section justifying the requirements based on previous studies and the current state of technology. Measurement procedures and test targets are provided where necessary. As in the previous chapter, measurement procedures have been drawn from several sources.[1–8]

## 8.1 Performance Requirements

Tables 8.1 and 8.2 provide size/resolution performance requirements for color and monochrome monitors. As was the case with luminance requirements, color and monochrome requirements differ somewhat because of the different levels of performance achieved with the two types of displays. Different applications may have different requirements; these differences are noted. In the tables, "H&V" refers to the horizontal and vertical dimensions.

**Table 8.1** Size/resolution requirements—Color.

| Measure | Minimum | Desired |
|---|---|---|
| Screen size (diagonal) | ≥ 17.5 in. | ≤ 24 in.[a] |
| Screen aspect ratio | Any | Same |
| Pixel aspect ratio | Square | Same |
| Addressability | ≥ 1280 × 1024 | ≥ 2048 × 1536 |
| Pixel density | ≥ 72 ppi | Same |
| Cm—Zone A | ≥ 35% H&V | ≥ 50% H&V |
| Cm—Zone B | ≥ 30% H&V | ≥ 50% H&V |

[a] Single-viewer applications and viewing distance of 18 in. (see justification in Sec. 8.3).

Table 8.2 Size/resolution requirements—Monochrome.

| Measure | Minimum | Desired |
| --- | --- | --- |
| Screen size (diagonal) | ≥ 17.5 in. | ≤ 24 in.[a] |
| Screen aspect ratio | Any | Same |
| Pixel aspect ratio | Square | Same |
| Addressability | ≥ 1280 × 1024 | ≥ 2048 × 1536 |
| Pixel density | ≥ 72 ppi | Same |
| Cm—Zone A | ≥ 35% H&V | ≥ 50% H&V |
| Cm—Zone B | ≥ 30% H&V | ≥ 50% H&V |

[a] Single-viewer applications and viewing distance of 18 in. (see justification in Sec. 8.3).

## 8.2 Measurement Definition

Screen size is the diagonal dimension of the viewable area as defined in Fig. 4.1. Screen aspect ratio is the width to height ratio of the viewable area (e.g., 4:3 or 16:9 for a display in landscape format where width is greater than height; the ratio is reversed for a display in portrait format). Pixel aspect ratio is the ratio of pixel height and width. Addressability is the number of individually commandable pixels. Pixel density is the number of pixels per inch in both the horizontal and vertical dimensions. Contrast modulation is as defined in Eq. (4.11) for a one-on/one-off grill pattern. Because performance on a CRT tends to degrade as the viewer moves away from the center of the monitor, two values are given. The first (Zone A) is defined as a circle with a diameter subtending 24 deg at an 18-in. viewing distance. This represents a 7.6-in. circle centered on the face of the monitor. Zone B is the remainder of the monitor excluding the outer 10% of the active screen area. Moiré is the artifact shown in Fig. 4.25 that results when phosphor pitch is large relative to pixel spacing. Pixel pitch should be ≤ 0.6 times the pixel spacing.

## 8.3 Requirement Rationale

For viewing most types of imagery, a nominal 20- to 21-in. diagonal display is generally considered a minimum. The viewable area is less than the nominal dimension; the 17.5-in. dimension represents the low end of the range for 20-in. CRT displays. For AMLCDs, the viewing area is about the same as the published diagonal. Thus, an 18-in. AMLCD is roughly equivalent to a 20-in. CRT. For a single viewer seated in front of a workstation display, there is a limit on display size before the viewer must move his or her head or body to view the full display. Viewing angles of greater than 30 deg generally require such movement. The 24-in. value shown allows a 30-deg display width at 18 in. for a 16:9 aspect ratio display. In theory, a 25-in. diagonal would meet the requirement for a 4:3 aspect ratio display.

For medical applications in particular, larger displays may be desirable for two reasons. First, they may allow larger images or more images to be displayed at one time. Second, they allow multiple viewers. Multiple displays, however, can accomplish the same purpose, possibly at lower cost. Any situation where the display is to be viewed at a distance greater than 16–20 in. may require a larger display with less pixel density (larger pixels).

If luminance and resolution measures are equal, there is no reason (other than cost) to prefer a 4:3 aspect ratio display over a 16:9 (and vice versa). Pixels (or pixel spacing), however, should be square in order to avoid distortion. Software can sometimes interact with monitor addressability to produce nonsquare pixel aspect ratios. Not all monitor pixel addressability ratios show the 4:3 aspect ratio. An addressability of 1280 × 1024, with the ratio of 4:3.2, is quite common.

Addressability and screen size combine to define pixel density. The key performance factor for many applications is pixel density because pixel density defines, for a given viewing distance, the visual angle subtended by the pixel (the smallest uniquely addressable element of the display). The smallest detail that can be seen, however, is a function of the HVS capability and the viewing distance. Maximum contrast discrimination is achieved when a pixel subtends about 1/4-deg of visual angle (2 cy/deg). Most current monitors have pixel densities of 72 ppi and greater. At a normal viewing distance of 16 in., this would equate to about 10 cy/deg. If we assume the viewer can move to within 10 in. of the display, it equates to 6.3 cy/deg. For a 100-ppi monitor, the equivalent frequency at 10 in. is 8.8 cy/deg, or roughly four times the frequency at which contrast sensitivity peaks.

In a study that measured Briggs target ratings and NIIRS performance, peak performance was achieved at a pixel density of 70–80 ppi for the NIIRS ratings.[9] Two monitors were used in the study, one with a pixel density of 170 ppi and the other with a pixel density of 100 ppi. The 100-ppi monitor was viewed at 1x and 2x magnification, and the 170-ppi monitor at 1x, 2x, and 3x (NIIRS ratings only). The 100-ppi monitor at 2x magnification was used as the standard for delta-NIIRS ratings. For the image data, bilinear interpolation was used. This has the effect of slightly degrading the resolution. Pixel replication was used for the Briggs targets. The NIIRS data are shown in Fig. 8.1. For the Briggs data, performance increased as pixels became larger (pixel density decreased). Results are shown in Fig. 8.2. In both figures, the square of the correlation coefficient is shown. The value $R^2$ indicates the proportion of variance in delta-NIIRS or Briggs (the dependent variable) accounted for by the log of pixel density (the independent variable).

In another study, imagery was displayed on both a monochrome and a color monitor at 1x and 2x magnification (bilinear interpolation); delta-NIIRS ratings were made relative to the hard-copy presentations of the same images.[10] The images were additionally displayed on the monochrome monitor with 2x pixel decimation (50 ppi). Pixel decimation (2x) is accomplished by deleting every other pixel in both the $x$ and $y$ dimensions. It does not preserve any information from the deleted pixels. Results are shown in Fig. 8.3. In this case, performance was maximized at 50 ppi. Note that the effect of magnification is greater for the color monitor when moving from 100 to 50 ppi.

The choice of pixel density is somewhat confounded by the effect of scene area and apparent sharpness. As pixel density increases, so does the amount of scene area that can

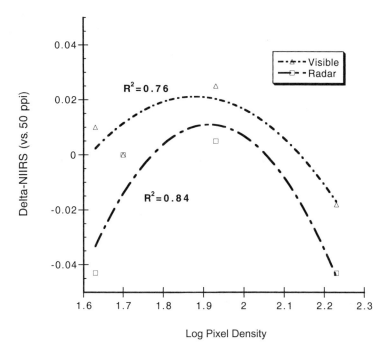

**Figure 8.1** Effect of pixel density on delta-NIIRS ratings (relative to 50 ppi).

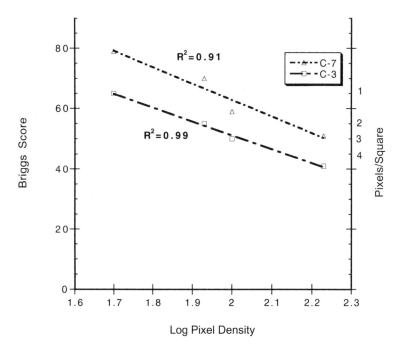

**Figure 8.2** Effect of pixel density on Briggs ratings.

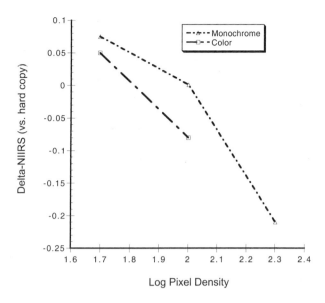

**Figure 8.3** Effect of pixel density on color and monochrome monitors.

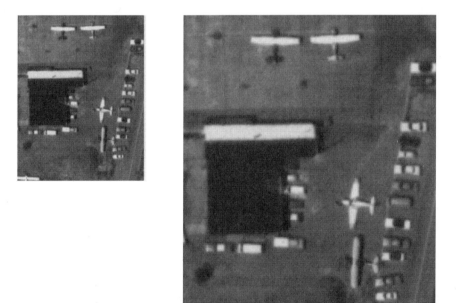

**Figure 8.4** Effect of pixel density on apparent sharpness.

be displayed at one time on the monitor. In addition, images appear sharper on monitors with higher pixel density. This is illustrated in Fig. 8.4. The larger-scale image appears less sharp than its smaller-scale counterpart. However, smaller detail can be resolved, since contrast sensitivity of the eye is better. The smaller scene detail is displayed at a larger visual angle where contrast sensitivity is improved (see Fig. 6.9). In addition, resolution with magnification is a function of the addressability and the interpolation used; resolution at maximum addressability is a function of spot size and shape.

The current tendency in display design is to increase pixel density to the maximum amount possible. Results of studies with aerial imagery indicate that information extraction performance is maximized between 70 and 80 ppi and is better at 50 ppi than 100 ppi.[9] For this and similar applications, higher densities do not appear to be necessary. For applications where the ability to see small, low-contrast detail is less important than portraying large numbers of pixels, higher densities may be desirable. Five-megapixel monitors with addressabilities of 2560 × 2048 may be useful for such applications, but no data as yet support such a need. Some AMLCDs with 9.2 megapixel addressability (> 200 ppi) have also been developed.[11] The same capability as a five-megapixel display can be achieved by displaying an image at 0.5x magnification on a monitor with 1280 × 1024 addressability.

The minimum addressability shown in Tables 8.1 and 8.2 provides pixel densities on the order of 66 to 86 ppi at the display sizes recommended, and 106 to 146 at the maximum addressability. At the latter pixel density, the desired range of 50 to 80 can be achieved with 2x magnification.

Contrast modulation (as used here) defines the ability of the monitor to portray high-contrast, fine detail. It is measured using grill patterns commanded to CLmax and CLmin. As the width of the bars in the grill become smaller, Lmax and Lmin are no longer achieved. Assuming Lmax/Lmin settings of 120 and 0.34 cd/m$^2$ respectively, a 50% Cm value is achieved when Lmax drops to 90 cd/m$^2$ and Lmin increases to 30.34 cd/m$^2$, both showing a 30 cd/m$^2$ change. Figure 8.5 plots the

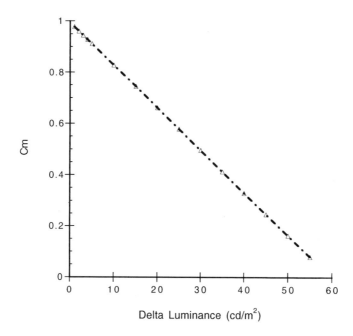

**Figure 8.5** Cm vs. luminance change in Lmax and Lmin.

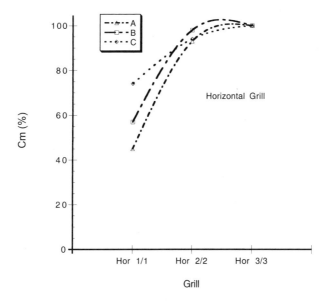

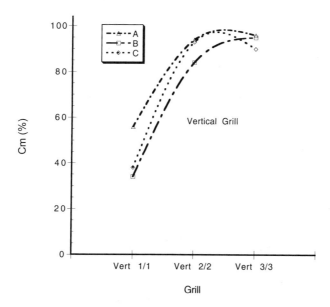

**Figure 8.6** Effect of grill size on Cm.

relationship. In fact, the decreases and increases do not necessarily balance because of factors such as halation and other nonuniform relationships between CL and luminance. Figure 8.6 shows actual measurement data on three color monitors.[12] The monitors were measured using horizontal and vertical grills with widths of

1 on/1 off, 2 on/2 off, and 3 on/3 off. Note that all of the monitors meet the Cm requirement with the 2 on/2 off grill. The Cm requirement can almost always be met with 2x magnification but at the expense of time, since four screens of data must be displayed instead of one.

The required level of Cm performance can be defined both theoretically and empirically. From a theoretical point of view, Cm must be above the detection threshold for the display spatial frequency. The Cm threshold for a 100-ppi monitor with luminance of 50 cd/m$^2$ is about 2.5% (see Fig. 6.10). Since this is the threshold for an average observer, a higher value is desirable. Data suggest a value of ~4% for the tenth-percentile observer.[13] If one were to display a square grid at Lmax/Lmin on a monitor with a Cm of 4%, 90% of the population would be expected to resolve the pattern. As luminance decreases, the required Cm value increases. For example, the threshold at a luminance of 1 cd/m$^2$ is on the order of 5%.[14] This would increase to about 10% for the tenth-percentile observer.

Rogers and Carel[15] indicate that Cm values should be 1.6 times the threshold values for "comfortable" detection. This would increase the 10% value to 16%. The NIDL requirements indicate a Cm value of 50% for text and 25% for imagery.[1] The rationale for the differing text and image values has not been explained by NIDL, but may be related to the higher autocorrelation function for imagery. Text has sharp on/off functions, so adjacent pixel values are not highly correlated. With continuous-tone imagery, tonal changes are more gradual so pixel values are more closely correlated.

In the empirical arena, two studies have investigated Cm thresholds, one directly and the other indirectly (with some confounding of the study variables). In the direct comparison, performance on two monochrome monitors was measured.[16] The two monitors were calibrated to a luminance range of 0.34 to 120 cd/m$^2$ (0.1 to 35 fL). Both had 100-ppi addressability and were viewed at both 1x and 2x magnification. Briggs ratings were made with the Briggs target centered on the monitor as well as placed in the upper left and lower right. Results are shown in Fig. 8.7. Performance was unaffected when Cm was above 50%. The NIIRS ratings showed no significant difference between the two monitors.

In the second study, a monochrome and a color monitor were compared with two viewing magnifications (1x and 2x with bilinear interpolation).[10] The monochrome monitor at 1x magnification had a Cm (geometric mean) of 86%, and the color monitor had a Cm of 26%. At 2x, the values were 99% and 88% respectively. At 1x magnification, the ratings on the monochrome monitor were significantly higher than those on the color monitor (Fig. 8.8). At 2x magnification, the differences were not statistically significant.

From these data, we can conclude that a Cm value of $\geq 50\%$ in the critical viewing angle (Zone A) is desirable. This level of performance is not achievable by all monitors. The minimum value of 35% reflects this. Currently there are color monitors on the market that meet the 50% goal; however, theoretical data suggest that the 35% value may be adequate for most of the population.

Although Chapter 4 defined several additional measures related to resolution, requirements for those parameters are not included here. Addressability and pixel density define pixel pitch; RAR and Cm are closely related, so only Cm is specified.

# SIZE AND RESOLUTION PERFORMANCE REQUIREMENTS

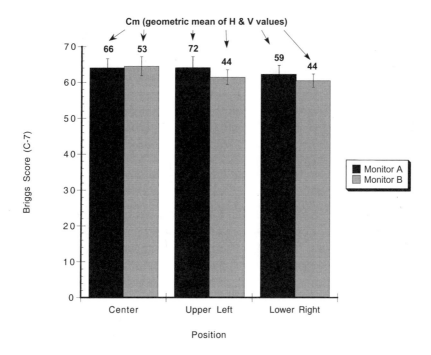

**Figure 8.7** Effect of Cm on Briggs scores. Values above bars are the geometric mean of horizontal (H) and vertical (V) Cm values.

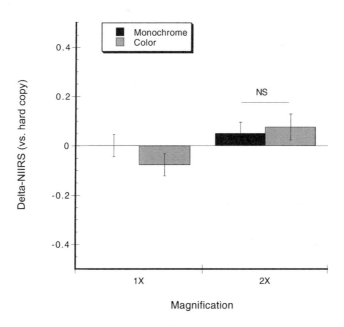

**Figure 8.8** Effect of monitor type and Cm on delta-NIIRS ratings.

In the case of RER, MTF, and edge sharpness, a sufficient basis for defining a minimum or desired value is not available at this time. The MTF of the image, the display, and the HVS together define the MTF of the viewing system. Logic would thus indicate that a high MTF and RER would be preferred. However, Eq. (4.12) demonstrates that RAR values of less than 1 produce both a high Cm value and a detectable raster structure. A definitive trade-off study has not yet been performed. Bandwidth affects horizontal resolution; the effect is captured in Cm measurements.

## 8.4 Instrument Measurement

In this section, a set of procedures for measuring size and resolution performance parameters is outlined. With the exception of Cm, size and resolution requirements can be verified with a tape measure and simple test targets. Measurement of Cm requires a scanning photometer or a CCD array. The initial setup prior to measurement should follow the procedures given in Sec. 7.4. Measurements should be made with the display calibrated to the Lmax and Lmin at which the display will normally be operated.

### *8.4.1 Screen size (diagonal)*

*Objective:* Define the size of the viewable screen.
*Equipment:* A flexible tape measure graduated in $1/16^{th}$-in. (or 1 mm) increments.
*Procedure:* Command a full screen image. Measure the size of the image diagonals (upper left to lower right, lower right to upper left).
*Analysis:* Average the two measurements to obtain the screen diagonal size.

### *8.4.2 Screen aspect ratio*

*Objective:* Define the screen aspect ratio as 4:3 or 16:9.
*Equipment:* A flexible tape measure graduated in $1/16^{th}$-in. (or 1 mm) increments.
*Procedure:* Command a full screen image. Measure the screen height at center. Measure the screen width at the center.
*Analysis:* Divide width by height for a landscape format monitor. Use the reverse for a portrait format monitor. For a landscape format monitor, a 4:3 aspect ratio screen gives a value of 1.33, and for a 16:9 ratio, a value of 1.77. Select the closest value. Reverse the ratios for a portrait format monitor.

### *8.4.3 Pixel aspect ratio*

*Objective:* Verify the pixel aspect ratio.
*Equipment:* A flexible tape measure graduated in $1/16^{th}$-in. (1 mm) increments.
*Procedure:* Command a test target of 400 pixels square with a CL corresponding to 50% Lmax. Center the target in a background of CL = 0. Measure the horizontal and vertical size of the box.

*Analysis*: Define the ratio of width to height. For a pixel density of ≤ 100 ppi, dimensions should be equal to within ±6%. For pixel densities >100 ppi, dimensions should agree to within 10%.

### 8.4.4 Addressability

*Objective*: Define the number of addressable pixels in the horizontal and vertical dimensions.
*Equipment*: A test pattern with pixels on 50% Lmax for the first and last addressable row and column and the two diagonals. The test pattern should also have a horizontal and vertical one on/one-off grill pattern.
*Procedure*: Display the test pattern. Confirm that the first and last row and column can be seen. Confirm there are no irregular jagged edges on the diagonals and no moiré on the grill pattern.
*Analysis*: If the test is passed, addressability is the number of pixels in the rows and columns. If the test fails, repeat it using a smaller pattern until the test is passed.

### 8.4.5 Pixel density

*Objective*: Determine the density of displayed pixels.
*Equipment*: A flexible tape measure graduated in $1/16^{th}$-in. (1 mm) increments and a test target with three equally spaced lines in the vertical and horizontal directions (Fig. 8.9).
*Procedure*: Measure the height of each vertical line and the width of each horizontal line. Average the measurements to determine the screen width and height of the active window.
*Analysis*: Divide the vertical screen height by the vertical addressability to define the vertical pixel density in ppi. Divide the horizontal screen width by the horizontal addressability to define the horizontal pixel density.

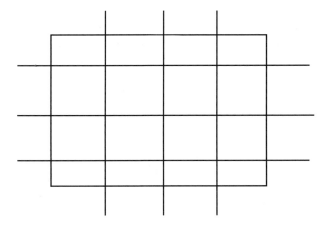

**Figure 8.9** Pixel density measurement.

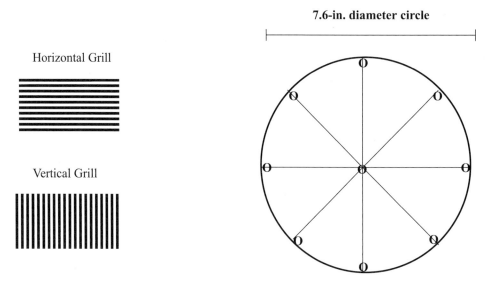

**Figure 8.10** Cm measurement—Zone A.

### *8.4.6 Contrast modulation—Zone A*

*Objective:* Define Cm in the central screen viewing area.
*Equipment:* Test targets consisting of vertical and horizontal grill patterns as shown in Fig. 8.10. One target is generated with the horizontal grill (one on/one off) positioned at the center of the circle and centered at the eight points on the circumference of the circle. The same procedure is repeated for the vertical grill pattern. Required equipment includes a scanning photometer or a matrix (CCD array) photometer.
*Procedure:* Display the vertical grill target and scan through the center of the target at the nine required locations. (With a matrix array, simply image the targets.) Repeat the procedure with the horizontal grill.
*Analysis:* For each of the nine positions, define the luminance of each line and space. Ignore the first and last line and space, and compute the average luminance for the lines and for the spaces. Define Cm for the horizontal and vertical grills at each location using Eq. (4.11). The lowest observed value is the Cm for Zone A.

### *8.4.7 Contrast modulation—Zone B*

*Objective:* Define Cm in the outer screen viewing area.
*Equipment:* Test targets consisting of vertical and horizontal grill patterns as shown in Fig. 8.10. One target is generated with the horizontal grill (one on/one off) positioned at eight points on the circumference of the

display as shown in Fig. 8.11. The same procedure is repeated for the vertical grill pattern. Required equipment includes a scanning photometer or a matrix (CCD array) photometer.

*Procedure:* Display the vertical grill target and scan through the center of the target at the eight required locations. (With a matrix array, simply image the targets.) Repeat the procedure with the horizontal grill.

*Analysis:* For each of the nine positions, define the luminance of each line and space. Ignore the first and last line and space, and compute the average luminance for the lines and for the spaces. Define Cm for the horizontal and vertical grill at each location using Eq. (4.11). The lowest observed value is the Cm for Zone B.

## 8.5 Measurement Alternatives

Most of the measurements relating to size and resolution are either provided by the vendor or can readily be made with a tape measure and calculator. The major exception, and probably the most important measurement, is Cm.

Vendors typically supply measures of screen size, addressability, and pixel pitch. Given size and addressability, pixel density can be calculated. As explained in Sec. 4.1.2, pixel pitch refers to the spacing between pixels and is simply 1 divided by the pixel density. However, for reasons unknown to the author, pixel density is typically expressed in inches and pixel pitch in millimeters. The pixel pitch measure must be described if it is to be useful. For example, for shadow mask monitors, diagonal pitch (the distance between two subpixels of the same color) is different than the horizontal pitch.

Contrast modulation measures may be made by a vendor, but they are seldom supplied. Instrumentation is quite costly and generally is beyond the budget of

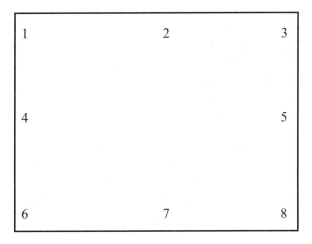

**Figure 8.11** Measurement points for Cm—Zone B. All points are 1 in. from the edge of the viewable area.

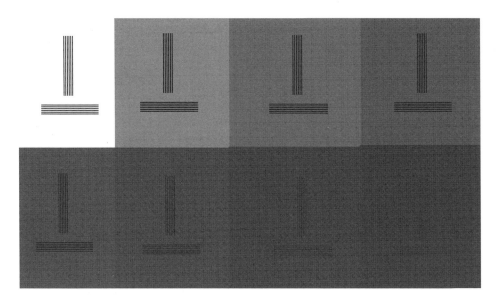

**Figure 8.12** Grill target for visual assessment of Cm.

most display buyers or users. As noted in Sec. 7.5, the NIDL is a source of these types of measures for a limited set of monitors on the market.

Five alternatives to the procedures described in the previous section are available. Perhaps the easiest to use is the Briggs target. Using the Briggs C-7 target on a properly calibrated monitor (see Chapter 10), the average score (with no magnification) should be between 55 and 65. This equates to the ability to see two (score of 55) or one (score of 65) pixel detail. On a 100-ppi monitor, the eye is a limiting factor, and many observers will not be able to see single pixel detail. If the monitor is not properly calibrated (see Chapter 10), readings may fall below the desired value for reasons other than Cm.

A second method uses the grill target shown in Fig. 8.12. The original targets show Cm values ranging from 0.99 to 0.075. For a monitor with a Cm performance of >0.50, all of the grills should be resolvable. As was the case with the Briggs target, the monitor must be properly calibrated.

In theory, one could also use a digital camera to find Cm values. The camera needs to be set up such that one monitor pixel is 8–10 times the size of a camera's pixel. The camera's contrast transfer function (CTF) should be defined and the camera then used to image the display. Alternatively, the camera's CTF can be ignored without too much concern and readings normalized to a three-on/three-off target. The concept is shown in Figs. 8.13 and 8.14.

Figure 8.13 shows screen shots of three targets. Figure 8.14 shows scans made across each of the targets. The scans were eight pixels wide. The modulation depth was measured for each of the three scans as shown. In CLs, the depths were 65, 64, and 45 for the 3/3, 2/2, and 1/1 targets, respectively. Translating the CLs to luminance levels and normalizing to the 3/3 target results in an MTF function as shown

# SIZE AND RESOLUTION PERFORMANCE REQUIREMENTS

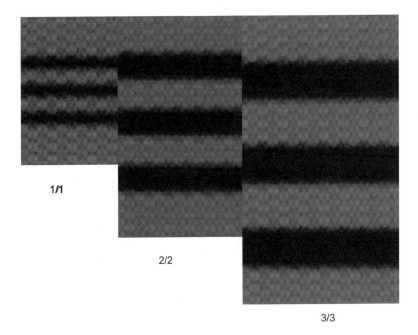

**Figure 8.13** Screen images of bar patterns on a shadow mask color monitor.

in Fig. 8.15. In this example, the monitor exceeds the Cm threshold of 50% with a value of 67%.

If one is willing to assume a Gaussian energy distribution in the pixel, a Cm value can be computed (given a knowledge of pixel size and pitch). An estimate of pixel size can be made directly with screen measurement or indirectly by imaging the screen. The width of a horizontal and a vertical line must be defined along with pixel pitch in the horizontal direction. For example, using the case of a one-bar grill shown in Fig. 8.13, the pixel width (width at 50% of the peak) is estimated at 1.04 pixels. The RAR is thus 1.04 but becomes 0.52 for alternating one-on/one-off patterns; thus,

$$RAR = \frac{W}{P}, \qquad (8.1)$$

where $W$ is the width of the pixel and $P$ is pixel pitch. In this case, the measurement is made in pixel space—normally it would be in English or metric distance. Software packages often measure in pixel space, since this measure is independent of addressability and display distortions and can be made in digital space. This was the case for the software used to generate Fig. 8.13. Measurement on the surface of the display would be required to accurately measure in English or metric distance units. Next, the following formula can be applied:

$$C_m = \frac{2}{\Pi} \exp[3.6(RAR) - 7(RAR^2) + (RAR)^3]. \qquad (8.2)$$

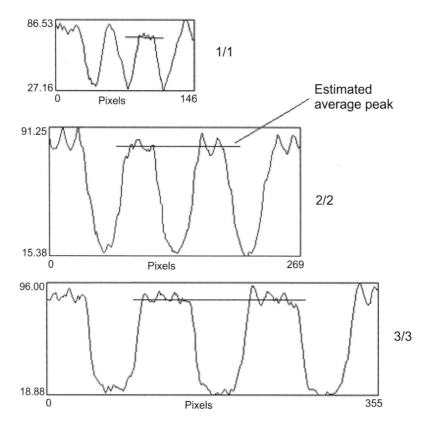

**Figure 8.14** Scans through bar targets.

This results in a Cm value of 71%, which is close to the value of 67% defined for the MTF function in Fig. 8.15.

Finally, if luminance can be measured, Cm can be estimated relative to a threshold value. A one-on/one-off grill pattern as shown in Fig. 8.16 is required. Each pattern should be larger than the aperture of the photometer. With a Cm of 1.0, the measured luminance of the grill will be 50% of the Lmax value (full field). As Cm decreases, the measured luminance of the grill decreases. If the ratio of grill luminance to Lmax is greater than 0.25, Cm is ~0.4. Thus, any ratio of > 0.25 can be considered adequate.

The measurements previously described were made using a square wave target. The MTF is normally defined using a sine wave target. Technically, we measured or computed a CTF with the square wave target. The MTF can be computed from the CTF using Fourier decomposition by

$$MTF_u = \frac{\pi}{4}\left[CTF_u + \frac{CTF_{(3u)}}{3} - \frac{CTF_{(5u)}}{5} + \frac{CTF_{(7u)}}{7} + \ldots\right]. \quad (8.3)$$

# SIZE AND RESOLUTION PERFORMANCE REQUIREMENTS

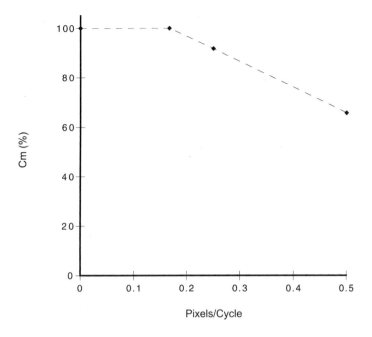

**Figure 8.15** The MTF for the display image in Fig. 8.12.

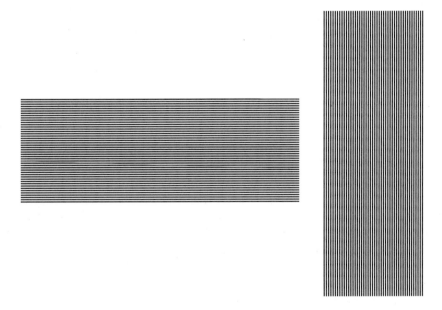

**Figure 8.16** Grill patterns for estimating Cm.

A square wave target will provide values greater than a sine wave target.[17] The difference is relatively small at the highest and lowest frequencies but can be as much as 20% greater for mid-frequencies.[8]

## 8.6 Summary

The ability to see fine detail on a monitor is a function of pixel size (pixel density) and Cm. Pixel density (ppi) is a function of addressability and the size of the active viewing area. The size of the active viewing area diagonal for most single-viewer applications should be within 17.5 (on a nominal 20-in. diagonal monitor) to 24 in. Smaller viewing sizes limit the size of the image that can be displayed and larger viewing sizes require movement of the viewer's head or body to see all of the display. Larger sizes can be used when multiple viewers are likely to use the display at one time or when viewing distance is greater than 18 in.

In order for a feature to be detectable, it must be of a size and contrast that can be resolved by the HVS. Many of the CRTs and AMLCDs on the market have pixel densities that are too fine to allow low-contrast detail to be resolved at a normal viewing distance. Studies suggest that a pixel density of 70–80 ppi is optimum for aerial imagery. The size and contrast of the image detail that must be resolved defines the optimum pixel density.

Contrast modulation in the context of display quality refers to the ability of the monitor to preserve and present fine detail. As detail becomes finer or smaller, contrast is lost and the detail may not have sufficient detail to be detectable. A Cm value of 50% or greater is considered adequate for viewing imagery.

With the exception of Cm, all of the size and resolution parameters can be measured without instrumentation other than a tape measure. Direct measurement of Cm requires a scanning or array photometer, both of which are quite expensive. Contrast modulation performance can be estimated using a digital camera or can be assessed using the Briggs target or grill targets on a calibrated monitor.

## References

[1] Video Electronic Standards Association, *Flat Panel Display Measurements Standard*, Version 2.0, VESA, Milpitas, CA (2001).
[2] National Information Display Laboratory, *Display Monitor Measurement Methods under discussion by EIA Committee JT-20, Part 1, Monochrome CRT Monitor Performance*, Version 2.0, Princeton, NJ (1995).
[3] National Information Display Laboratory, *Display Monitor Measurement Methods under discussion by EIA, Part 2, Color CRT Monitor Performance*, Version 2.0, Princeton, NJ (1995).
[4] National Information Display Laboratory, *Test Procedures for Evaluation of CRT Display Monitors*, Version 2.0, Princeton, NJ (1991).
[5] J. Leachtenauer, *SC Measurement Procedures*, NIMA, Reston, VA (1998).

[6] National Information Display Laboratory, *Request for Evaluation Monitors for the National Imagery and Mapping Agency (NIMA) Integrated Exploitation Capability (IEC)*, Princeton, NJ (1999).

[7] H. Roehrig, "The monochrome cathode ray tube display and its performance," in Y. Kim and S. Horii (eds.), *Handbook of Medical Imaging, Vol. 3, Display and PACS*, SPIE Press, Bellingham, WA, pp. 157–220 (2000).

[8] P. A. Keller, *Electronic Display Measurement*, John Wiley & Sons, Inc., New York (1997).

[9] J. Leachtenauer, A. Biache, and G. Garney, " Effects of pixel density on softcopy image interpretability," *Final Program and Proceedings, PICS Conference*, Society for Imaging Science and Technology, Savannah, GA, April 25–28, 1999, pp. 184–188.

[10] J. C. Leachtenauer and N. L. Salvaggio, "NIIRS prediction: use of the Briggs target," *ASPRS/ASCM Annual Convention and Exhibition Technical Papers, Vol. 1, Remote Sensing and Photogrammetry*, American Society for Photogrammetry and Remote Sensing and American Congress on Surveying and Mapping, Baltimore, MD, April 22–25, 1996, pp. 282–291.

[11] Y. Hosoya and S. L. Wright, " High-resolution LCD technologies for the IBM T220/T221 monitor," *Society for Information Display, 2002 International Symposium, Digest of Technical Papers*, Vol. XXXIII, No. 1, pp. 83–85 (2002).

[12] J. Leachtenauer, A. Biache, and G. Garney, *Effect of Contrast Modulation on Softcopy Image Interpretability*, Technology Assessment Office, National Imagery and Mapping Agency, Reston, VA (1998).

[13] R. J. Farrell and J. M. Booth, *Design Handbook for Imagery Interpretation Equipment*, Boeing Aerospace Co., Seattle, WA (1984).

[14] A. van Meeteran and J. Vos, "Resolution and contrast sensitivity at low luminances," *Vision Research*, Vol. 12, pp. 825–833 (1972).

[15] J G. Rogers and W. L. Carel, *Development of Design Criteria for Sensor Displays*, Hughes Aircraft Company, Culver City, CA (1973).

[16] J. Leachtenauer, G. Garney, and A. Biache, "Contrast modulation—how much is enough?" *Final Program and Proceedings, PICS Conference, Society for Imaging Science and Technology*, Portland, OR, March 26–29, 2000, pp. 130–134.

[17] J. R. Schott, *Remote Sensing: The Image Chain Approach*, Oxford University Press, New York (1997).

# Chapter 9
# Noise, Artifact, and Distortion Performance Requirements

This final chapter on display requirements considers requirements for noise, artifacts, and distortions. Noise represents unwanted luminance variation and can be defined in both the spatial and temporal domain. Artifacts can be considered a form of noise in that they interfere with the presentation of the image of interest. Distortions result from failures to geometrically control the spatial relationships in the image of interest. Noise and artifacts in particular can affect the ability to see small, low-contrast detail. Noise is of particular importance in the display of medical images that require the detection of small anomalies (tumors, fractures). Distortions are of lesser importance, primarily because they are so readily detectable and correctable. As in the previous two chapters, this chapter begins with a listing of requirements and their definitions. Justification for the requirements follows, along with procedures for measurement.[1-8] Test targets are provided on the enclosed CD.

## 9.1 Performance Requirements

The requirements for noise, artifacts, and distortion are listed in Table 9.1. With the exception of moiré, these requirements apply to both color and monochrome displays. The extinction ratio applies to stereo displays, and pixel defects apply to AMLCD and plasma displays. The SNR is listed as an important parameter without a defined minimum value. It is difficult to specify the SNR because the many contributors of noise may not be separable. The SNR desired value is based on an information theory calculation.

## 9.2 Measurement Definition

In Table 9.1, warm-up time is the time required for the monitor CL/luminance relationship to stabilize—specifically, the time required for Lmax to stabilize. The refresh rate is the cyclic rate at which each pixel is refreshed; a rate of 72 Hz indicates the display is refreshed 72 times/second. The minimum and desired rate applies to CRTs. Because the light source remains activated in AMLCDs, lower rates on the order of 55 Hz can be tolerated. However, when rates fall below the minimum value, the monitor may appear to flicker.

Jitter, swim, and drift represent movement of the raster pattern over different periods of time ranging from 0.5 to 60 seconds. Jitter is also used to describe

**Table 9.1** Noise, artifact, and distortion performance requirements.

| Measure | Minimum | Desired |
|---|---|---|
| Warm-up time | 30 minutes ±50%, 60 minutes ±10% | 10 minutes ± 10% |
| Refresh rate | ≥ 72 Hz[a] | ≥ 85 Hz |
| Jitter | < 2 mils (0.002 in) | Same |
| Swim/drift | < 5 mils | Same |
| Macro/micro jitter | None apparent | None apparent |
| Luminance step response | No ringing | Same |
| Moiré[b] | None | Same |
| Extinction ratio | ≥ 20:1 monochrome/15:1 color | As high as possible |
| Mura/artifacts | Undefined | None detectable |
| Pixel defects | 0.01% | 0.001% |
| SNR | Undefined | ≥ 48 dB[c] |
| Straightness | < 0.5% | Same |
| Linearity | < 1.0% | Same |

[a] Monoscopic viewing on a CRT. See discussion for AMLCDs in Secs. 9.2 and 9.3. For stereo viewing, 60 Hz per eye (120 Hz total) is the minimum.

[b] Color displays only.

[c] For an 8-bit display.

movement when image translation is occurring. In so-called image roam, the image is translated across the display. If the translation rate exceeds the capability of the hardware/software suite to keep pace with the translation, the image may appear to move in large jumps (macro jitter) or in small vibratory motions (micro jitter). In the case of macro jitter, the image is generally uninterpretable during the motion and the jitter is, of course, annoying to the viewer. Micro jitter has the effect of blurring information in the image and thus reduces resolution. It is also annoying.

Luminance step response is the luminance response of the monitor across a dark to light or light to dark edge. A phenomenon called "ringing" can occur when the electronics overcompensate for the change (see Fig. 4.24). Moiré (Fig. 4.25) results in a banding pattern and occurs on displays where the mask to pixel spacing (pixel pitch) equals or exceeds 1.0. The extinction ratio measures the degree to which the image for one eye is obscured from the other eye. Mura is a term that refers to a variety of artifacts resulting from imperfections or nonuniformities in the display; it tends to be used more for AMLCD displays. "Pixel defects" is a term used to quantify nonoperating pixels in an AMLCD or plasma display. Individual pixels may fail to respond (partly or totally) to commanded changes.

Signal-to-noise ratio is the ratio of mean signal (luminance) to the standard deviation of signal around the mean. Several of the measures listed contribute to the perception of noise (unwanted signal variation). They include the luminance variations that occur before a monitor is warmed up; the refresh rate (flicker resulting from low rates); jitter, swim, and drift; luminance step response; the extinction ratio (for stereo displays); and pixel defects (AMLCD/plasma). Pixel defects can also be considered as artifacts. Under some circumstances, the mask or grill used in color displays can be considered a source of noise since the mask or grill is always at a very low (or no) luminance level regardless of signal level. Finally, an RAR of less than 1 produces noise at any combination of contrast and viewing distance where pixels can be resolved.

Straightness and linearity refer to the geometric pattern of the raster. Linearity is the degree to which horizontal and vertical raster patterns are parallel. Straightness is the degree to which horizontal and vertical lines maintain linearity.

## 9.3 Requirement Rationale

The minimum value listed for warm-up time is based on current CRT technology; AMLCD devices have much shorter warm-up times. Before a monitor has warmed up to at least ±50% of Lmax, it will probably be in violation of several other performance parameters such as Cm and DR. Variations in Lmax during warm-up as high as 30% have been observed. Note that so-called power saver modes may affect warm-up performance.

The threshold value for the refresh rate (72 Hz) is based on the desire to avoid perceptible flicker. As discussed in Sec. 4.3, not all individuals perceive flicker at the same rate. Further, the perception of flicker depends on the screen or scene luminance level and is more apparent at high levels of luminance. It would thus tend to be more noticeable on a chest x-ray as opposed to an MR image, or on a snow scene as opposed to an urban scene. With AMLCDs, a lower rate (perhaps to 55 Hz) can be tolerated since the LCD backlighting is always on. A performance-based value has yet to be defined. In the case of stereo displays, achieving the desired rate of 72 Hz per eye requires a refresh rate of 144 Hz. This rate is generally not available with commercial displays, so the requirement (as specified in Table 9.1) was lowered to 120 Hz. Note that reducing addressability often allows higher refresh rates; thus, the addressability for a stereo display can be traded for increased refresh rate.

Values for jitter, swim, and drift are based on what is routinely achieved with current technology. Larger values could probably be tolerated with little impact on performance. Jitter during image translation can, of course, most easily be avoided by slowing the rate of translation. Where the task requires a search of images larger than the display addressability, image roaming at a relatively rapid rate may be required. Certain real-time reconnaissance systems require the observer to monitor a scene that moves at the same rate (scaled to the display) as the aircraft. Thus, the imaging system in the aircraft images the ground at the same rate as the aircraft flies. Similarly, video systems with rapid translation require image motion at a rapid rate. Controllers with very fast response times have been developed for use with computer action games. However, other aspects of image quality may suffer with such controllers.

As with flicker, ringing (Fig. 4.21) and moiré (Fig. 4.22) should be avoided. Moiré occurs when a repetitive pattern is displayed with a frequency near that of

**Figure 9.1** Effect of less than complete stereo extinction.

the mask spacing. Ringing occurs when the response to a CL change in the scan direction successively overshoots and undershoots the target luminance level. Both moiré and ringing produce unwanted contrast changes in an image. Ringing also reduces resolution.

The extinction level values are technology-driven. Ideally, there should be no residual image as the two eye views alternate in stereo viewing. At present, phosphor, refresh rate, and viewing device limitations preclude the ability to achieve total extinction. This results in a loss of contrast or increase in noise; the effect has yet to be parametrically determined. Figure 9.1 is a simulated example in which the right image is not fully extinguished so a low-contrast offset image is overlaid on the other member of the stereo pair. Note the blurring in the right half of the figure.

As explained in Sec. 4.4, mura describes a variety of blemishes that can be seen on the face of an LCD monitor when the monitor is driven to a constant level.[1] The blemishes result from factors such as nonuniformities in liquid crystal distribution and impurities in the liquid crystal material. The blemishes appear as low-contrast brightness nonuniformities that may have regular or irregular shapes. Figure 9.2 shows some examples. Mura-like (i.e., fixed pattern) blemishes can also occur in CRTs.[1]

"Artifact" is a term used more specifically to describe such things as ghosting and streaking. "Ghosting," as the name implies, refers to a ghost image displaced in the scan direction from the original image. Ghosting can occur with incorrect cable connector matching or excessive cable lengths. Figure 9.3 provides an example.

In an ideal world there would be no mura or artifacts, but this is not the case. Mura and artifacts that are detectable and occur randomly generally do not present a viewing problem since they can be separated from the image of interest. Artifacts such as ghosting and streaking do create problems because they reduce contrast

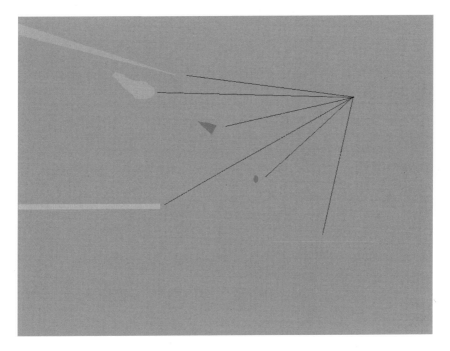

**Figure 9.2** Examples of mura.

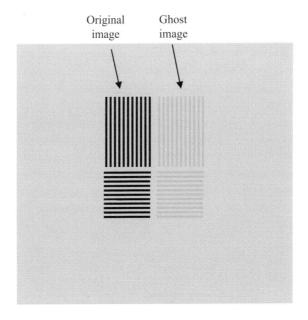

**Figure 9.3** Ghosting.

**Figure 9.4** Pixel defects at a 0.01% rate.

and/or add noise. Therefore, they should be eliminated or minimized. Artifacts that may be confused with targets of interest are a particular problem. In a monitor comparison study, three experienced radiologists indicated that artifacts of 0.35 mm and greater were detectable on a chest radiograph displayed on a CRT, but that artifacts of 0.25 mm and smaller were not. The radiologists agreed that a 0.45-mm artifact was potentially distracting.[9] Plasma display panels and AMLCDs may have individual pixel defects or even line defects where the pixel or line remains in a constant state regardless of the intended luminance level. The minimum and desired values shown in Table 9.1 are based on what has been achieved in technology, not on measured task performance. Figure 9.4 shows pixel errors at the minimum rate (0.01% of total pixels or 17 errors for the image shown). The lines point to sample errors, but the errors seem to present little distraction.

Up to this point, noise and artifact parameters have been discussed in terms of measurable display attributes. The tendency has been, however, for task performance to be specified in terms of the SNR as opposed to the measurable display parameters. As noted previously, several of the display measures can contribute to noise.

Noise can be characterized in both the spatial and temporal domain. An area on a display commanded to the same level will vary in luminance as a function of both position and time. Noise can also be characterized as temporal or frequency-dependent ("pink" noise) or spatial or time-independent ("white" noise). The SNR is the ratio of the average signal level to the standard deviation of the signal level.

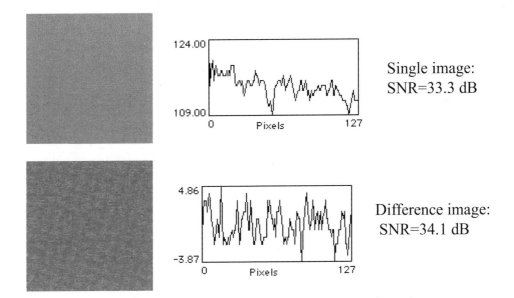

**Figure 9.5** SNR measurements.

The general effect of noise is to decrease the detectable Cm threshold. Small contrast differences become more difficult or impossible to detect. Figure 9.5 shows an example of SNR measured across a flat-field single image of a display and a difference image obtained by subtracting two images obtained at different points in time. Note that the single field image shows both a high- and low-frequency variation. The low-frequency variation is the result of luminance non-uniformity. In both cases, the high-frequency variation is of about the same magnitude.

For a display, SNR varies as a function of CL or luminance; SNR is higher at high luminance/command values. A difficulty in interpreting the results of previous studies on SNR arises from the fact that measurement procedures are often not fully defined. For example, SNR may be defined in terms of drive voltage, which does not take into account phosphor variations.

One method of addressing the effect of noise is in terms of information theory. The SNR should be sufficient to avoid a loss in bit rate as defined by

$$I = \frac{\log_{10}[1 + SNR]}{\log_{10} 2}. \tag{9.1}$$

For an 8-bit display, the SNR should be > 255:1 or 48.1 dB. For a variety of reasons, display systems do not generally achieve this level of performance. Values of 25 to 35 dB are more typical. At a value of 35 dB, the theoretical number of bits of information is reduced to 5.8. However, the visual system integrates over time and space so the net loss is less than suggested by the theoretical bit loss.

Because of the various contributors to the SNR, no specific minimum SNR requirement can be defined. The Barten model discussed in Chapter 6 contains a noise term, but the noise is image noise rather than display noise. The "noiseless" model actually includes the noise of the displays used in developing the model. Aside from the parameters specified elsewhere (e.g., luminance uniformity, jitter, swim, and drift), the two major contributors to display noise are the video controller noise and phosphor nonuniformity.

The effect of phosphor SNR differences (P-45 vs. P-104) was investigated in a study where radiologists searched for pulmonary nodules on computed radiography (CR) images.[10] Untrained observers also evaluated the effects of the phosphor differences in terms of the number of discriminable gray levels. The P-45 phosphor had a better SNR level than the P-104 phosphor and showed significantly better gray-level discrimination. No statistically significant difference in nodule detection performance was observed in an ROC analysis. In another study, the nose power spectra (NPS) of four monitors was measured, one with a P-104 phosphor and three with a P-45.[11] The noise power was highest for the P-104 and one of the P-45 devices. The power spectra agreed with the subjective impressions of noise. No explanation for the high-noise P-45 device was provided.

The specified values for straightness and linearity are based on what is generally achievable with current technology. Lack of straightness on a large scale results in such things as keystone and trapezoid distortion. These distortions are normally adjustable with monitor controls. Small-scale straightness errors can distort normally straight features such as letters, building edges, etc. Although such distortions may be annoying, they generally do not affect information extraction performance. Nonlinearity errors produce pixel size differences as well as distortions.

## 9.4 Instrument Measurement

Measurement of noise, artifact, and distortion performance requires the use of the greatest diversity of measurement instrumentation: a timing device, a calibrated CCD array (or a scanning photometer in one case), a photometer, a magnifier and a loupe with a 0.001-in. scale, and various test targets. However, most of the distortions and artifacts can be assessed visually to determine at least their potential impact. As in previous sections, the initial setup should follow the procedures given in Sec. 7.4, and the monitor should be calibrated to its intended values of Lmax and Lmin.

### 9.4.1 Warm-up time

*Objective:* Define compliance with warm-up time specification.
*Equipment:* A clock and a photometer that can read from 0.03 cd/m$^2$ (0.01 fL) to 686 cd/m$^2$ (200 fL) with accuracy of ±5%.
*Procedure:* Follow the set-up procedure in Sec. 7.4. Measure Lmax every 5 minutes for 1 hour. Ensure that the screen saver and power saver modes are disabled.

*Analysis:* Plot CLmax luminance readings and determine when the readings are constant to within ±10%. Readings should be within 50% of the desired Lmax at 30 minutes and within 10% after 60 minutes.

### 9.4.2 Scan rate

Scan rate is provided by the manufacturer and normally does not need to be measured unless visual inspection (i.e., flicker) contradicts the manufacturer's data. Note that scan rate often varies as a function of addressability. See the NIDL document in Ref. [6] for measurement procedures.

### 9.4.3 Jitter, swim, and drift

*Objective:* Measure variation in beam spot position as a function of time.

*Equipment:* A spatially calibrated CCD array with resolution and positional stability equivalent to 10% of pixel size. A grill pattern as shown in Fig. 9.6 containing three lines that are one pixel wide, with the middle line centered in the display and the outer lines located a distance equal to 5% of the screen dimension (active area) from the edge of the active area. A timing device with 0.1-sec precision.

*Procedure:* Place the vertical grill at the center of the display and measure the position of the center line for 150 sec at a rate equivalent to the refresh rate (e.g., 60 times per sec for a 60-Hz rate). Repeat for the horizontal grill. Repeat the process at the top and bottom corners of the display. Measure vertical and horizontal motion of the display/measuring device by attaching a horizontal and vertical knife edge to the center of the display and measuring positional variation over a 30-sec period. Variability must be less than 0.1 mil (0.001 in.).

*Analysis:* Define the average position of each grill line over the measurement period. Define maximum deviation from mean deviation for each grill pattern at 0.5 sec (jitter), 10 sec (swim), and 60 sec (drift). Report maximum deviation over the five measurement positions.

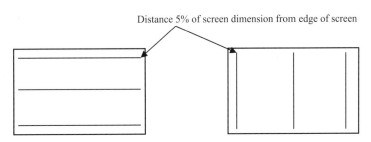

**Figure 9.6** Jitter/swim/drift target.

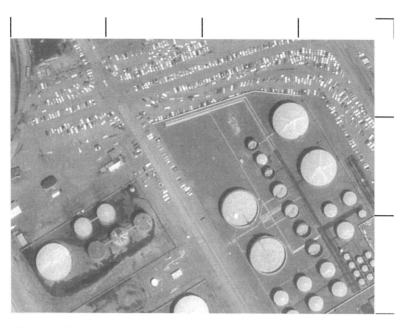

**Figure 9.7** Micro and macro jitter targets (image and calibration marks).

## 9.4.4 Macro and micro jitter

*Objective:* Determine the presence of macro and micro jitter.
*Equipment:* A timing device with 1-sec precision. Test targets as shown in Fig. 9.7. One test target will divide the screen into $n$ equal intervals, where $n$ is the screen addressability divided by a roam rate (pixels/second). For example, with a screen addressability of 768 × 1024, splitting the screen into four intervals allows for testing at a 256-pixel/second roam rate. The vertical dimension would be split into three intervals to test the same rate. The second test target is any image of interest that is at least twice as large as the screen addressability (e.g., 2048 × 1536 for a screen with 1024 × 768 addressability).
*Procedure:* Move the image over the display in the horizontal and vertical direction at a defined translation rate that is a multiple of the addressability. Look for evidence of movement at a frequency higher than the translation rate. With micro jitter, the image will appear to vibrate. With macro jitter, the image will first remain stationary and will then jump to a new position.
*Analysis:* Define the slowest translation rate at which jitter occurs.

## 9.4.5 Luminance step response

*Objective:* Determine the presence of ringing resulting from overshoot/undershoot.

# NOISE, ARTIFACT, AND DISTORTION PERFORMANCE REQUIREMENTS

**Figure 9.8** Luminance step response measurement.

*Equipment:* A scanning photometer or CCD array that can read from 0.03 cd/m$^2$ (0.01 fL) to 686 cd/m$^2$ (200 fL) with accuracy of ±5%.

*Procedure:* Define a test target with a box 15% of screen size centered in a full-screen display. The box should be commanded to 90% of Lmax and the surround to 10% of Lmax. Display the target and measure positive and negative luminance transitions along three equally spaced scan lines through the box as shown in Fig. 9.8.

*Analysis:* Transitions within the box should be ≤ 5% of Lmax. Transitions in the background should be ≤ 10% of Lmax.

## 9.4.6 Moiré

*Objective:* Determine lack of moiré.

*Equipment:* A loupe with a scale graduated in 0.001 in. or equivalent.

*Procedure:* Measure the phosphor pitch in the vertical and horizontal dimensions at the screen center. For aperture grill screens, vertical pitch will be 0. Define pixel size by 1/pixel density.

*Analysis:* Define the value of the phosphor pitch/pixel size. A value of ≤ 0.6 passes.

## 9.4.7 Extinction ratio

*Objective:* Measure the stereo extinction ratio.
*Equipment:* A photometer that can read from 0.03 cd/m$^2$ (0.01 fL) to 686 cd/m$^2$ (200 fL) with accuracy of ±5%. Two "stereo" pairs of images with full addressability. One pair has the left image at CL 255 (or CLmax) and the right image at 0. The other pair has the right image at CL 255 (or CLmax) and the left image at 0.
*Procedure:* Calibrate the monitor to 0.34 cd/m$^2$ (0.1 fL) Lmin and 120 cd/m$^2$ (35 fL Lmax) with no room lighting. Measure the ratio of Lmax to Lmin on both the left- and right-side images through the stereo system. Measure at the center of the screen and at the eight points defined in Fig. 8.11.
*Analysis:* Extinction ratio (left) = L (left, on, white/black)/(left, off, black/white).
Extinction ratio (right) = R (right, on, white/black)/(right, off, black/white).

## 9.4.8 Mura and other artifacts

*Objective:* Determine the presence of mura and other artifacts and catalog them.
*Equipment:* Test targets consisting of a flat field at 50% of Lmax and a grill target as shown in Fig. 9.9.
*Procedure:* Display the flat-field target and look for any anomalies (areas of the screen having different luminance levels such as those shown in Fig. 9.2). Record their positions and measure luminance levels. Display the grill target and look for ghosting or streaking.
*Analysis:* Mura should not be noticeable at a normal viewing distance. Any ghosting or streaking is probably evidence of a cabling/connection problem and should be corrected.

**Figure 9.9** Grill target.

## 9.4.9 Pixel defects

*Objective:* Determine the pixel defect frequency (rate).
*Equipment:* Test targets consisting of full-frame flat fields at Lmax and Lmin.
*Procedure:* Display the flat fields and count the number of pixels that are stuck-on (white in the Lmin field) or stuck-off (black in the Lmax field). Use a magnifier to view the images.
*Analysis:* Divide the total number of defective pixels by the total number of pixels and report the result as a percentage.

## 9.4.10 Signal-to-noise ratio

*Objective:* Define display SNR.
*Equipment:* A test target consisting of a full image at Lmax. A CCD array with a known SNR.
*Procedure:* Image the display at a magnification such that one pixel of the display covers four pixels of the array. Integrate (expose) the display over the maximum time possible (at least twice the refresh rate and preferably several times the rate). More accurate methods are provided in Refs. [2] and [7].
*Analysis:* Normalize the image such that the maximum digital value is at the display maximum. Compute the mean and standard deviation. Define SNR as

$$SNR = \frac{S_{avg}}{\sigma_{display}}, \qquad (9.2)$$

where $S_{avg}$ is the average signal level and $\sigma_{display}$ is the standard deviation corrected for the camera's standard deviation. This is calculated using

$$\sigma_{display} = \sqrt{\sigma_{total}^2 - \sigma_{camera}^2}. \qquad (9.3)$$

This method does not account for temporal variations.

## 9.4.11 Straightness (waviness)

*Objective:* Define nonlinearities in a raster pattern.
*Equipment:* A spatially calibrated CCD array with resolution and positional stability equivalent to 10% of pixel size. An x-y translation stage with 0.1-mm accuracy. A grill pattern as shown in Fig. 9.6; the grill should be at 100% of Lmax.

*Procedure:* Display the horizontal grill and define the *x-y* position of the center. Measure deviations from the center coordinates for each of three grill lines at intervals corresponding to 5% of the addressable screen width. Repeat for the vertical grill.

*Analysis:* Tabulate deviations in *y* for the horizontal grill and *x* for the vertical grill. Report the maximum deviation.

### 9.4.12 Linearity

*Objective:* Define raster nonlinearity.
*Equipment:* A target as shown in Fig. 9.10. A CCD array and a calibrated *x-y* translation stage.
*Procedure:* Use a detector to locate the center of the vertical lines at the horizontal center line of the display. Measure the *x* positions of each vertical line along the horizontal center line. Repeat for the horizontal grill, measuring the *y* position of each line along the vertical center line of the display.
*Analysis:* Define the average spacing between the vertical lines. Define the deviation from this average for each vertical measurement point (*x* spacing – average *x* spacing) as a percentage of the average. Repeat for the horizontal dimension (*y* spacing – average *y* spacing). Report the maximum percentage delta.

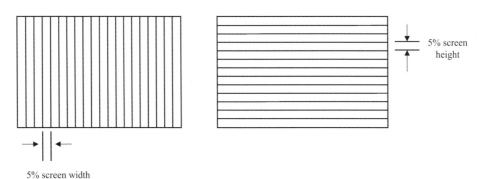

**Figure 9.10** Linearity target.

## 9.5 Measurement Alternatives

Distortions and artifacts can often be observed without instrumentation, thus requiring a judgement call as to the severity of the effect. A light meter can be used to assess warm-up time following the procedures given in Sec. 7.5 by commanding the screen to Lmax and defining an exposure value (EV). A change of one EV represents a 100% increase or 50% decrease. This is obviously a crude approximation. A better estimate can be obtained by imaging the screen commanded to CLmax with a digital camera at a constant EV and then measuring the average CL of the same area in each image. An area measurement is required to average over positional and temporal variations in output luminance. Extinction ratio could theoretically be measured in a similar fashion, although with questionable accuracy.

The presence of jitter, swim, and drift can be assessed by displaying a grill pattern and observing it at 2x optical magnification or higher. Performance at the specified levels would not be detectable, but substantially poorer performance (on the order of 20 mils) would be observed. This suggests that the specified values of 2 and 5 mils are not detectable with normal viewing.

Luminance step response can be assessed with the target shown in Fig. 9.8. The center area is 15% of the screen height commanded to 90% of Lmax, and the surround is at 10% of Lmax. This can be roughly approximated (in the absence of a photometer) by setting the center area to CL 250 and the surround to CL 120. The vertical edges of the box should be inspected for evidence of ringing. Alternatively, a digital camera can be used to photograph the target (Fig. 9.11). An intensity scan across the target can be used to determine the presence of ringing.

Moiré, if present, is observable in a flat field. It can also be evaluated through measurement with a magnified reticle graduated in .001-in. or 0.025 mm.

Signal-to-noise ratio can be estimated with a digital camera. For a monochrome display, the screen should be commanded to Lmax and imaged such that at

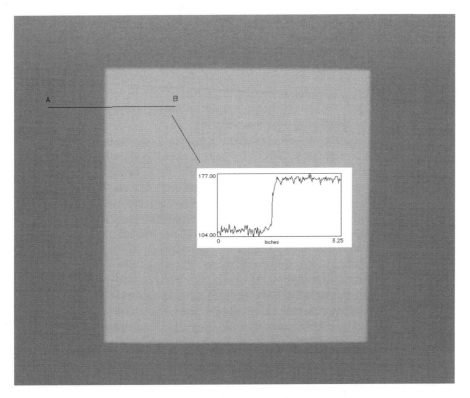

**Figure 9.11** Digital image of step response target with histogram of edge AB.

least four camera pixels cover one cycle (two pixels) on the monitor. Exposure time should be set so as to average over at least two refresh cycles (e.g., 1/30$^{th}$ of a second for a 60-Hz refresh rate). With a 100-ppi monitor, a cycle covers 0.02 in. With a 1600 × 1200 pixel camera, four pixels should cover 0.02 in. In this example, the camera should be set up such that the horizontal camera field of view (FOV) covers 8 in. on the display. The formula is

$$FOV_{display} = \frac{CD_{display}}{4} \times Horizontal\ Pixels_{camera}, \qquad (9.4)$$

where $FOV_{display}$ is the width of the camera FOV on the display, $CD_{display}$ is the width (cycle distance) of a cycle on the display, and $Horizontal\ Pixels_{camera}$ is the number of camera pixels in the horizontal dimension.

To estimate spatial SNR, an image is acquired and displayed. The image is transformed through a level adjustment such that CLmax is 255 (for an 8-bit system). A histogram of a 400 × 400 pixel square is obtained and the standard deviation defined. Although several software packages have the capability of defining the statistics of an image histogram, NIH Image and Image J are freeware packages that offer the capability. Both are available from www.nih.org.

The SNR is defined as the ratio of the histogram mean to the measured standard deviation. An estimate of temporal variation can be obtained by acquiring multiple images over time and determining the standard deviation over difference images (as shown in Fig. 9.5). Both values should be in the region of 50 to 300 (34 to 50 dB) where

$$SNR_{dB} = 20\log_{10}SNR. \qquad (9.5)$$

A color display separates the three colors with a mask or grill. Thus, at the micro level, the color monitor has very low SNR since the display alternates between high-luminance color subpixels and black masks or grills. When the display is viewed, however, the eye integrates over the subpixels and mask. If the camera technique is used, it may be necessary to undersample slightly by placing three camera pixels over four display pixels.

A relatively crude estimate of the noise power spectrum (NPS) can be obtained by imaging the display and taking an FFT of the resultant image. Averaging several radial traces through the center of the FFT image provides the NPS. Reference [7] provides details.

Finally, straightness and linearity can be assessed visually using the targets shown in Figs. 9.6 and 9.11. However, errors of the magnitude specified are not likely to be detectable.

## 9.6 Summary

A variety of factors contribute to noise. Noise reduces the ability to see fine detail and low-contrast differences. Noise may occur in both the spatial and temporal domain. Temporal noise includes fluctuations in the beam position (jitter, swim, and drift) as well as fluctuations related to electronic warm-up. Spatial noise includes variations related to voltage variations and phosphor and LCD response. Artifacts such as pixel defects, mura, ringing, moiré, and such effects as ghosting and streaking can also be considered as forms of noise. In general, the tendency has been to characterize and specify various contributors to noise as opposed to specifying the SNR directly. With a few exceptions, the noise/artifact parameters that contribute to the SNR are generally well controlled with current displays. Artifacts such as ghosting and streaking are the exception; they occur when monitors are incorrectly set up (see Chapter 10). Straightness and linearity measure the geometric properties of the display. Straightness covers both small- and large-scale departures from geometric uniformity; the large-scale distortions are generally adjustable with monitor controls. Small-scale distortions and linearity are generally well controlled with current displays to the point where they are not major quality factors.

## References

[1] Video Electronic Standards Association, *Flat Panel Display Measurements Standard*, Version 2.0, VESA, Milpitas, CA (2001).
[2] National Information Display Laboratory, *Display Monitor Measurement Methods under discussion by EIA Committee JT-20, Part 1, Monochrome CRT Monitor Performance,* Version 2.0, Princeton, NJ (1995).
[3] National Information Display Laboratory, *Display Monitor Measurement Methods under discussion by EIA, Part 2, Color CRT Monitor Performance,* Version 2.0, Princeton, NJ (1995).
[4] National Information Display Laboratory, *Test Procedures for Evaluation of CRT Display Monitors,* Version 2.0, Princeton, NJ (1991).
[5] J. Leachtenauer, *SC Measurement Procedures,* NIMA, Reston, VA (1998).
[6] National Information Display Laboratory, *Request for Evaluation Monitors for the National Imagery and Mapping Agency (NIMA) Integrated Exploitation Capability (IEC)* Princeton, NJ (1999).
[7] H. Roehrig, "The monochrome cathode ray tube display and its performance," in Y. Kim and S. Horii (eds.), *Handbook of Medical Imaging, Vol. 3, Display and PACS,* SPIE Press, Bellingham, WA, pp. 157–220 (2000).
[8] P. A. Keller, *Electronic Display Measurement*, John Wiley & Sons, Inc., New York (1997).
[9] J. F. Copeland, et al., "Practical quality control standards for digital display monitor*s,*" *Proc. SPIE*, Vol. 3976, pp. 315–322 (2000).

[10] E. Krupinski and H. Roehrig, "Observer performance using monitors with different phosphors—an ROC study," *Proc. SPIE*, Vol. 4324, pp. 18–22, (2001).
[11] M. Weibrecht, G. Spekowius, P. Quadfleig, and H. Blume, "Image quality assessment of monochrome monitors for medical softcopy display," *Proc. SPIE*, Vol. 3031, pp. 232–244 (1997).

# Chapter 10
# Monitor Selection and Setup

The previous three chapters discussed factors affecting the quality of displayed images and provided recommended levels for each of several parameters. No single display system necessarily meets all of the desired performance levels, and not all of the performance parameters are equally important. This chapter begins with guidance on monitor and video controller selection, followed by a section on how a display should be set up in terms of environmental controls and perceptual linearization of the monitor response.

## 10.1 Monitor and Video Controller Selection

In selecting a monitor and video controller, several decisions must be made as a starting point. The first is the issue of color vs. monochrome monitors. Monochrome imagery can be viewed on color monitors, but color imagery cannot be viewed in color on monochrome monitors. The color will appear as shades of gray that may be a function of all three colors (RGB) or simply one color (e.g., green). Annotation capability is also limited on monochrome monitors; they are not a good choice for a multiple-purpose monitor, i.e., for viewing imagery and for office applications. Table 10.1 lists issues that affect monitor selection plus recommendations. Monochrome monitors tend to have better Cm performance at 1x magnification. The difference generally disappears at 2x and higher magnification. If imagery is to be viewed at 2x magnification or higher in order to resolve all of the image detail, Cm performance on both color and monochrome CRT monitors will be about equal. Monochrome CRT monitors tend to have better theoretical SNR performance, since no mask or grill is present. Color monitors without masks have not yet reached the commercial market. Because of the Cm and SNR performance of monochrome CRT monitors at 1x magnification, they are generally preferred for primary diagnosis in medical applications, particularly when the task is performed at 1x magnification. The same can be said for image search in military applications. The significant luminance loss occurring with the use of stereo viewing favors monochrome (as opposed to color) CRT monitors, provided that Cm performance can be maintained at high luminance levels. An alternative is a color AMLCD with high luminance levels. Monochrome monitors do not have a multisync capability, meaning that they are very sensitive to timing issues and a given monitor may not correctly interface to the controller. Finally, monochrome CRTs cost on the order of 3 to 8 times the cost of color CRTs, and longevity has been an issue.

**Table 10.1** Color vs. monochrome monitor selection issues.

| Issue | Response | Recommendation |
|---|---|---|
| Imagery type | Color | Color |
| | Monochrome | Color or monochrome |
| Magnification | None | Monochrome |
| | 2x or higher | Color or monochrome |
| Application | Primary diagnosis | Monochrome |
| | Secondary/other | Color or monochrome |
| | Stereo | Monochrome |
| Cost | Limited budget | Color |
| | Not a significant issue | Monochrome or color |

A second issue to be addressed is the type of monitor–CRT, AMLCD, or PDP. Issues and recommendations are provided in Table 10.2. An AMLCD monitor weighs less and occupies less space than a CRT monitor; CRTs are sensitive to magnetic fields (because of the magnetics used to control beam deflection) so AMLCDs (or PDPs) may be preferred in strong magnetic environments. The luminance output of an AMLCD is a function of the backlighting intensity, and increased luminance does not degrade resolution as it does with CRTs. In at least one instance, an AMLCD operated in a relatively bright environment (190 lx or 18 fc) performed as well as a CRT in an 11-lx (1-fc) environment.[1] Plasma displays can be very large and thus suitable for group viewing, although projection CRT or LCD displays are viable alternatives. Plasma displays, and to a lesser degree, AMLCDs, tend to have relatively slow response times and thus may not be well suited for applications such as video display or those requiring rapid roam (such as search of large images). Finally, color CRTs are significantly less costly than AMLCDs, at least in terms of the initial purchase price. Monochrome AMLCDs may prove to be cost competitive with monochrome CRTs because of the ability to simply replace the backlighting system when it fails.

**Table 10.2** Monitor type selection issues.

| Issue | Response | Recommendation |
|---|---|---|
| Viewing environment | Limited space/weight | AMLCD |
| | Strong magnetic fields | AMLCD/PDP |
| | Bright room lighting | AMLCD |
| Application | Group viewing | PDP |
| | Video | CRT |
| Cost | Limited budget | CRT |
| | Not a significant issue | CRT or AMLCD |

Having determined the type of display, one must examine performance trade-offs. Tables 7.1 through 7.3, 8.1, 8.2, and 9.1 provided minimum and desired performance levels. At a minimum, a display system should show $\geq$ 25.4-dB DR, an

Lmax of 120 cd/m$^2$ (35 fL) or higher (170 cd/m$^2$ or 50 fL for a medical monitor), ≥ 50% Cm (at least in the center of the display), a bit depth of at least 8 (10 for medical radiography applications), and the maximum possible SNR. There should be no noticeable artifacts or distortions such as ghosting, streaking, moiré, ringing, flicker (due to low refresh rate), etc. For specific applications, other factors such as stereo extinction ratio and roam rate without jitter need to be considered. Addressability may be of concern for some applications, but high addressability may not be usable if pixel density is high. Finally—although not a monitor characteristic per se—the display system should have the capability to be calibrated to a defined Lmin and Lmax and have a perceptual linearization LUT applied.

A final system element to be considered is the controller. In many display systems, the controller is transparent and is simply a part of the operating hardware. For those cases where workstations are assembled from components, several factors should be considered in the selection of the controller. The card must be matched to the monitor in terms of addressability, refresh rate, impedance, interface requirements, and the platform on which the board will operate. Two key performance parameters are the bit depth of the board and the presence of a calibration capability. Controllers are currently designed to operate at 8, 10, or 12 bits. As will be discussed in a later section, the process of perceptual linearization reduces the number of unique CLs. In 8-bit space, on the order of 20% of the CLs can be lost due to quantization. Quantization occurs when, in the process of defining the perceptual linearization LUT, multiple-input CLs result in the same output level (or adjacent input values go to separated output levels). Key performance parameters are the bit depth of the controller and the presence of a calibration capability. If the images to be viewed are 8-bit images, defining the LUT in 10-bit space reduces the loss of CLs. For images at higher bit levels, developing the LUT in 12 bits before displaying an image at 10 bits reduces or eliminates quantization error.

Some graphics cards or controllers are furnished with software and associated measurement hardware that can be used to calibrate the output of the display in terms of Lmax, Lmin, and the total I/O function. Boards are available on the medical market, for example, that provide the capability to calibrate a monitor in accordance with the NEMA/DICOM standard. Other desirable attributes include high clock frequency (>360 MHz) and separate horizontal and vertical sync.[2] For medical applications, the ability to support a portrait format is usually a requirement. Many of the measures previously discussed for displays, such as Cm, temporal stability, SNR, and edge response, apply to controllers as well. The type of graphics card/monitor interface—digital vs. analog—is also a consideration. Most monitors now employ a digital brightness and contrast control. Application-specific integrated circuits (ASICs) and microprocessors bridge the gap between pure analog and pure digital controllers. Although digital controllers are used for AMLCDs and PDPs, an advantage for CRTs has not yet been shown.[3]

As indicated earlier, very little of the desired display performance information is available from vendors. For government users, the NIDL is a source of data. Their web site is at http://www.nidl.org/accomp_mstr.htm., but reports cannot be accessed without proper authorization. Another source of useful data is at http://

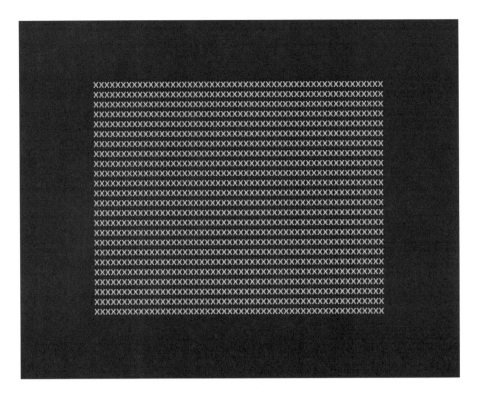

**Figure 10.1** Focus target.

www.sid.org/displaytechnologies/display.html. A web site can be found for most display and controller vendors by searching for the company name. Current monochrome vendors include Clinton Electronics, Dome, MegaScan, Pixellink, and Siemens. Color display vendors include Barco, Cornerstone, EIZO, Hitachi, IBM, Mitsubishi, Nokia, Samsung, Sony, and Viewsonic. In many cases, manuals for the displays can be found on the web at the company's site. Additional sources may be found in Refs. [4] and [5].

If reports are not available from the NIDL, a set of test targets can be used to make at least a preliminary assessment of monitor or monitor/controller quality. Every attempt should be made to determine Lmax, either from the vendor or from measurement. Contrast modulation performance can be assessed using the Briggs target (C-7) or the Cm targets shown as Fig. 8.12 and included on the enclosed CD. The target should be viewed at both the center and the corners of the monitor. The target shown in Fig. 10.1 is useful in assessing focus over the full display. The target described in Ref. [6] can be used to evaluate viewing angle effects for color LCD monitors. Finally, the SMPTE target can be used to evaluate general display quality.[7] Once a preliminary assessment has been made, the more detailed procedures described in Chapters 7–9 can be applied.

## 10.2 Monitor Setup

A newly purchased monitor must be set up before use. Monitor setup consists of four steps:

1. Connecting the monitor to the CPU and setting the timing/addressability;

2. Preparing the monitor environment;

3. Setting the monitor calibration; and

4. Developing and installing the perceptual linearization LUT.

Each of these steps is discussed in the following sections.

### *10.2.1 Monitor connection and setup*

For individual PC user setups, connecting the monitor is straightforward and simply requires installing the vendor-supplied cables and following the vendor's instructions. Note that some video boards or controllers may conflict with monitor settings. If a nonstandard controller has been installed, timing conflicts (refresh rates/addressability) need to be resolved.

For more complex setups where multiple monitors or workstations may be connected to a central CPU, the situation is more complex. In this type of environment, cabling may be lengthy and signal losses may occur. Impedance needs to be matched between the driver and display and the cabling length minimized. Impedance is a measure of electrical current resistance expressed in ohms ($\Omega$). If impedance is not matched (e.g., 75 $\Omega$) throughout a transmission system, error-producing reflections can be produced. If the cabling cannot be shortened, a video amplifier may be needed. A digital interface can minimize this problem.

Once the monitor is connected, flat field gray-level and grill targets should be displayed to look for any evidence of display artifacts related to the signal transmission from the CPU to the monitor. Any evidence of ghosting and streaking is cause for concern and should be investigated and eliminated. Color temperature should be set and a geometric pattern displayed to check for and adjust any raster distortions such as barrel or pincushion. The viewing area should be centered and the size adjusted (for a CRT). Any other settings should be adjusted according to vendor specification except for brightness and contrast. These controls will be adjusted in the calibration process.

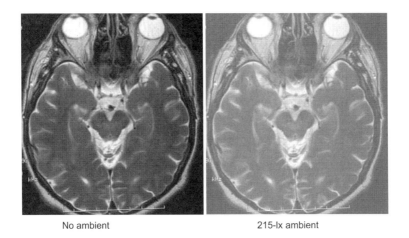

No ambient            215-lx ambient

**Figure 10.2** Effect of ambient light on an MRI display.

## *10.2.2 Controlling the monitor environment*

The most important factor in the monitor environment is lighting around the display (called ambient light). As noted previously, any light falling on the display adds to Lmin and thus decreases DR. Ideally, the monitor should be located in a room where ambient light can be reduced to ~11–22 lx (1–2 fc) in the vicinity of the monitor. Ideally the increase in measured Lmin resulting from room lighting should be less than 10% of the desired Lmin. Figure 10.2 illustrates the effect of allowing ambient light to fall on an MRI display. Figure 10.3 shows the loss in gray

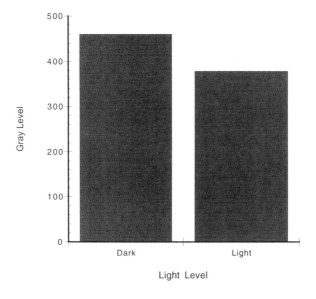

**Figure 10.3** Effect of 215-lx room lighting on the number of gray levels.

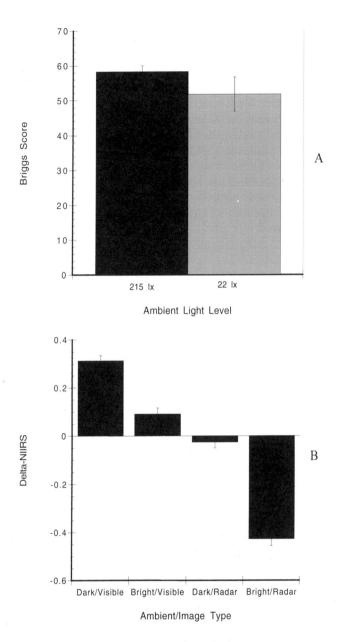

**Figure 10.4** Effect of 215-lx (bright) vs. 22-lx (dark) ambient light level on Briggs (C-7) and NIIRS ratings.

levels (based on the Barten model) when room lighting is increased to a 215-lx level and 3.1 cd/m$^2$ is added to the output luminance without changing the display calibration (0.34–120 cd/m$^2$).

The effect of ambient light has been investigated in several previous studies. Figure 10.4 shows the effect of a 215-lx vs. a 22-lx lighting environment on NIIRS

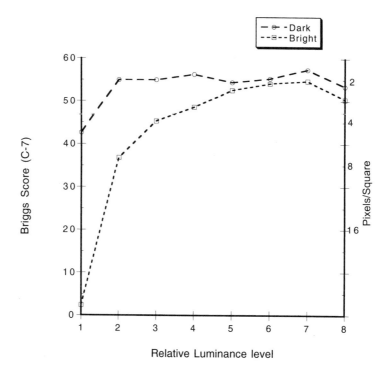

**Figure 10.5** Effect of 215-lx vs. 22-lx ambient light on Briggs scores as a function of the luminance level.

ratings and Briggs target ratings.[8] In the case of the Briggs ratings, display performance of the darkest targets suffers (Fig. 10.5).

The NEMA/DICOM calibration procedure recommends that the display be measured in the presence of the intended light level, and the perceptual linearization function adjusted to take into account the ambient light. However, since ambient light adds to the display luminance, the net result is still a loss in DR. The DR of a display can be increased by increasing Lmax. However, with a CRT, it is generally not possible to increase Lmax to the point where DR can be maintained. Adding 3.1 cd/m$^2$ to a display calibrated to 0.34–119 cd/m$^2$ (25.4-dB DR) would require increasing Lmax to 1204 cd/m$^2$ in order to maintain the same DR. A color CRT is generally limited to an Lmax of about 120 cd/m$^2$ (35 fL) and a monochrome display seldom runs above 685 cd/m$^2$ (200 fL). An AMLCD can be driven to high levels but may appear too bright at these levels depending on the image displayed.[1] Figure 10.6 shows the effect of attempting to overcome a 215-lx ambient light level for both visible spectrum and radar imagery. With the same Lmin and Lmax (0.34 and 120 cd/m$^2$) as for the dark (11 lx) environment, the NIIRS loss is 0.2 for visible and 0.4 for radar. Decreasing Lmin and increasing Lmax reduces but does not overcome the loss.[8]

The impact of ambient light is so important in the medical community that the ACR has defined lighting standards for mammographic reading rooms.[9] The lumi-

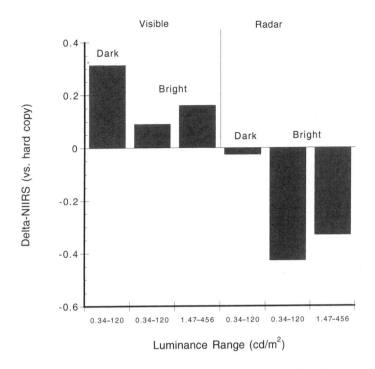

**Figure 10.6** Effect of increasing Lmin and Lmax on 215-lx lighting.

nance output of CRTs in particular is substantially less than light boxes, so any loss of DR due to ambient light can be particularly injurious. Control of ambient lighting is recognized as a significant issue in picture archiving and communication systems (PACS).[10,11]

Under some circumstances it may not be possible to reduce ambient light to the desired level, but it is still possible to reduce the light falling on the face of the display. This can be accomplished by some combination of monitor placement relative to the light sources and shielding of the light sources or the monitor. In a typical office with overhead fluorescent fixtures, the best solution is to turn off the overheads and replace them with desk lamps or track lighting that illuminates the wall behind the display. Window lighting should be eliminated if at all possible The face of a monitor should never be directed toward a light source, nor should a monitor be positioned directly under or be illuminated by an overhead light. If lighting is reduced in an area but an adjacent room or hall is brightly lit, some means of avoiding sudden increases in light level when the door to the adjacent room is opened should be taken. The effects of overhead lighting can be reduced by installing directional diffusers in the lights. Black egg crate diffusers substantially reduce light except for directly below the source. Even a simple black cardboard shield around the monitor face can be of benefit if no other solution is available.

Figure 10.7 compares the performance of a black egg crate diffuser with that of the standard white prismatic diffuser.[12] The figure shows relative luminance inten-

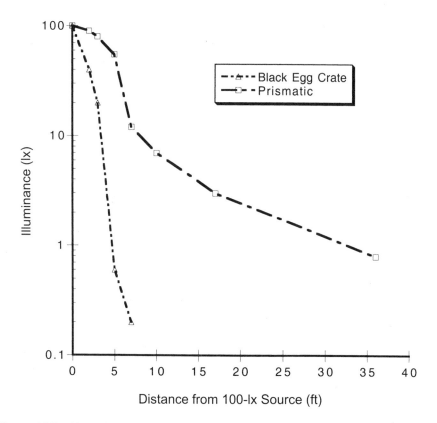

**Figure 10.7** Effect of diffuser type on relative luminous intensity. Data from Ref. [12].

sity on the face of the monitor assuming the light source (a fluorescent fixture perpendicular to the monitor) is six feet above the monitor and the monitor face is perpendicular to the floor.

The Briggs target can be used as a simple test of the success of ambient light control. The C-3 and C-7 targets should be read with no ambient light present, then read again in the intended lighting environment. A decrease of more than three score levels for the darker targets is cause for concern.

Although little or no light should fall on the face of a monitor, the room should not be in total darkness. Ideally, the monitor surround (background) should be about equal to the average monitor luminance[13] so the viewer's pupils do not need to adapt when looking beyond the boundaries of the display. This can be accomplished with track lighting or even with strategically placed desk lamps.

In an environment where hard copy material must be referenced at the same time a display is viewed, the material should be placed such that its illumination does not fall on the face of the monitor. Careful placement of the material and its light source or shielding the monitor from the light source can eliminate the problem.

Cathode-ray tubes are particularly susceptible to strong magnetic fields. A magnetic current is used to deflect the electron beam that produces the image on

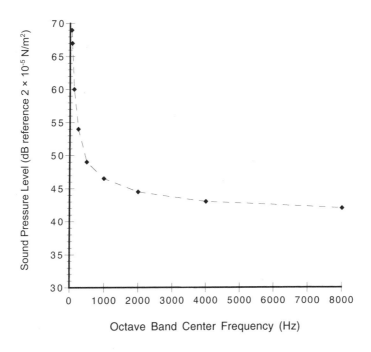

**Figure 10.8** NC-45 noise criterion. Data from Ref. [12].

the display. Even other monitors may affect a display. Moving a monitor relative to another monitor or even rotating a monitor may change displayed colors—in extreme cases, even the CL/luminance function. Some type of shielding should be employed to reduce the problem. A change in colors/luminance levels or evidence of noise in a flat field is an indication of magnetic interference. Cathode-ray tubes have a degaussing control that affects color purity; the monitor should be degaussed if any evidence of color impurity is evident.

Finally, high levels of acoustical noise and poor workstation configuration can reduce the performance obtained from a workstation that may otherwise be of high quality. The NC-45 criterion has been recommended for display rooms.[12] This criterion defines allowable sound pressure levels as a function of frequency. Figure 10.8 plots the criterion limit. Ergonomic considerations dictate, at a minimum, an adjustable chair, adjustable input device (keyboard, mouse) height, and adjustable monitor position and tilt. Although it is not uncommon to see, both the monitor and input device should not be placed on the top of a desk and the user required to cope with an uncomfortable work environment.

### *10.2.3 Monitor calibration*

Once the monitor is set up, it needs to be calibrated. Some graphics cards perform this function by requiring the user to simply place a measurement device on the face of the monitor and follow a set of measurement and control input procedures.

The monitor is set to a predefined Lmin and Lmax, and a perceptual linearization table calculated and applied. Where such a capability does not exist, it is necessary to measure the luminance of the monitor as a function of input CL. This is a two-step process. First, Lmin and Lmax are adjusted to predetermined values (e.g., 0.34 and 120 cd/m$^2$) using a full-field display commanded to CLmin and CLmax. This is typically an iterative process, since changing one value may change the other. It may also be necessary to change color gain settings or even addressability in order to achieve the desired Lmax. Note that calibration should be performed under the intended room lighting conditions. Once the Lmin and Lmax values have been set, intermediate points must be measured following the procedure in Sec. 7.4 for initial setup. At this point, the perceptual linearization function can be selected and calculated.

### *10.2.4 Perceptual linearization*

Perceptual linearization applies a LUT such that the difference between any two output luminance levels is perceptually equal. Figure 10.9 shows a set of measured data (observed) and the desired relationship to achieve perceptual linearization. A LUT is required that modifies the CL/luminance relationship in order to produce the desired function. For example, a CL of 150 was observed to produce an output luminance of ~ 14 fL (48 cd/m$^2$). The desired output is ~8 fL (27 cd/m$^2$). On the observed function, the desired output is achieved with a CL of 110. The LUT thus transforms all CLs of 150 to a level of 110. A similar transform is made for every original CL. Figure 10.10 shows a typical LUT relationship where input is the original CL and output is the transformed value necessary to achieve perceptual linearization.

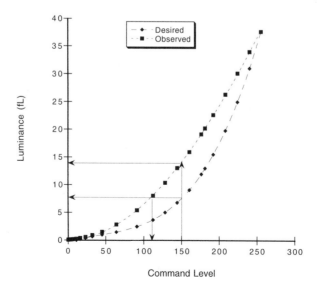

**Figure 10.9** Example of observed and desired CL/luminance functions.

# MONITOR SELECTION AND SETUP

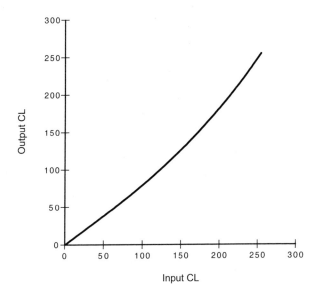

**Figure 10.10** Perceptual linearization LUT function.

The process of creating a LUT requires knowledge of the relationship between CL and luminance for both the observed and the desired functions. In other words,

$$CL = f(\text{Luminance}). \tag{10.1}$$

Some type of curve-fitting routine is required to develop the relationship.

The general form of an observed CL/luminance function is described by

$$L = a + bCL^c. \tag{10.2}$$

It can also be approximated by

$$L = L_{min} + (L_{max} - L_{min})v^\gamma, \tag{10.3}$$

where $L_{min}$ and $L_{max}$ are the minimum and maximum luminance values, $v$ is the normalized digital-to-analog converter (DAC) value ($L_{min}$ occurs at DAC = 0 and $L_{max}$ at DAC = 1), and $\gamma$ is the monitor gamma (the slope of the CL/luminance function in log/log space).

Solving the function in Eq. (10.2) (or its inverse to predict CL from L) requires a nonlinear curve-fitting capability. Alternatively, a third- or fourth-order polynomial can be used to solve for CL such that

$$CL = a + bL + cL^2 + dL^3. \tag{10.4}$$

The correlation between CL and luminance should be ≥ 0.999. Given the two equations to predict the CL from the observed and desired luminance values, the two equations can be run against the same set of luminance values (either the desired or observed or even equally spaced values between Lmin and Lmax) so that each of the two equations is used to predict CL values from the same set of luminance values. At this point we have two sets of command values, the observed and the desired. From Fig. 10.9, we observe a CL of 150 but desire 110 to achieve the desired luminance level. We thus regress the desired CL values onto the observed values such that (typically)

$$CL_{desired} = a + bCL_{observed} + cCL_{observed}^2 + dCL_{observed}^3. \qquad (10.5)$$

A fourth-order polynomial may be required and some degree of adjustment may be necessary at very low or high values to avoid out-of-range predictions (<0, >255 for an 8-bit system). The resultant LUT is applied as the last step before displaying an image.

What may not be immediately apparent is that the implementation of a perceptual linearization LUT results in a loss of gray levels due to quantization if the LUT bit depth is the same as the image and display. If the number of input and output CLs available is the same (and both are integer values), some loss will occur given a nonuniform transform. The loss is typically on the order of 20% of the original gray levels. Performing the linearization at a higher bit depth avoids the problem. For example, DACs may be operated at 10 bits before displaying an image at 8 bits.

The effect of a perceptual linearization LUT has been examined in several studies. The medical community was instrumental in developing and adopting the NEMA/DICOM standard.[13] A study of lesion detection on mammograms showed that detection performance was significantly higher on a perceptually linearized display;[14] total viewing times and dwell times were shorter on the linearized display, as well. Other studies have shown that the number of discernable gray levels increases with perceptual linearization. A study using multispectral imagery showed a 0.05 NIIRS gain using the NEMA/DICOM function over the standard CRT power law function.[15] The difference for Briggs scores was not statistically significant. A comparison of the NEMA/DICOM,[13] Crites,[16] and Rogers and Carel[17] linearization functions showed the latter to be inferior to the former two for radar and visible monochrome imagery.[18] The Crites and NEMA/DICOM functions did not differ significantly.

A comparison of the Crites, NEMA/DICOM, and Rogers and Carel functions is shown in Fig. 10.11 in terms of the relationship between normalized CL and Cm thresholds. The current Barten model[19] gives yet a slightly different function. For medical applications, the NEMA/DICOM function is the standard, although the revised Barten model should perhaps be examined as a replacement for some applications. For other applications, the Barten model appears to be the best choice because of its generality. It is recommended that the noise submodel be implemented based on image noise statistics, and also that other submodels be specific

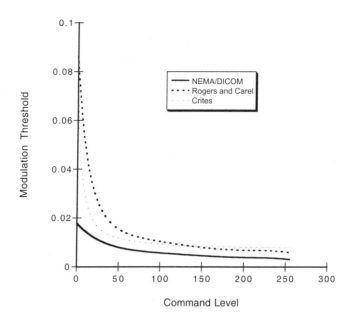

**Figure 10.11** Modulation thresholds for referenced models.

for the intended viewing conditions and tasks. This would include specific provision for the display spatial frequency under the intended viewing conditions, the angular size of the target, and the phosphor type.

In Chapter 6, a correction factor to the CSF for coherently illuminated imagery was provided, and it was suggested that this correction might be appropriate for radar imagery. Equation (6.16) provides an adjustment to a CSF calculated using one of the perceptual linearization models. The author has not evaluated this approach with radar imagery, but it may be useful to try at some point.

The process of constructing the desired CL/luminance function involves successive computations of modulation thresholds, each one defined as a luminance change and added to the previous luminance level. For example, assume an Lmin of 0.34 cd/m$^2$ (0.1 fL). If the modulation threshold ($M_t$) is 0.02 (considering all of the other conditions affecting $M_t$), then the desired luminance at the next CL (1) is 0.353 (0.104 fL). If, at this level, the threshold is still 0.02, then the next desired level is 0.367 cd/m$^2$. The process is repeated successively until some desired Lmax value is reached. The NEMA/DICOM function covers the range of 0.05cd/m$^2$ (0.015 fL) to 3986 cd/m$^2$ (1163 fL).[13] The Crites function covers the range of 0.1 to 48 fL.[16] The number of steps between the desired Lmin and Lmax will not equal the number of CLs, so normalization must be performed. For example, the NEMA/DICOM function shows 465 steps between an Lmin of 0.34 cd/m$^2$ and an Lmax of 120 cd/m$^2$. For a 10-bit system, the 465 steps must be spread across 1024 levels. The linear equation to translate the 465 NEMA/DICOM steps to the CL values of 0–1023 is

$$CL = -2.2047 + 2.2047ND, \qquad (10.6)$$

where ND is the NEMA/DICOM step number (1—465). It is evident that the relationship is approximate since CL values must be integer values.

It is convenient to describe the cascaded luminance values computed from a perceptual linearization function as an equation. The equation for the NEMA/DICOM function was given previously as Eq. (6.13). A closed form approximation has been defined as[20]

$$y = (e^{ax+b} - c)^d, \quad (10.7)$$

where

> $x$ = NEMA/DICOM index variable ($i$ = 1–1023),
> $y$ = luminance in fL,
> $a$ = 0.00325935271299,
> $b$ = 0.475628811499555,
> $c$ = 1.51, and
> $d$ = 1.87.

The IDEX function uses a polynomial to describe the relationship between normalized luminance and CL. The equation can thus be applied to any range of CLs and a luminance range of 0.1 to 48 fL. The equation for a range of 0.1 to 35 fL is defined as

$$y = 0.1 + 4.060264x - 6.226862x^2 + 48.145864x^3 - 60.928632x^4 + 49.848766x^5, \quad (10.8)$$

where

> $0 \leq x \leq 1.0$ ($x$ is the normalized CL) and
> $0.1 \leq y \leq 35$ ($y$ is the desired luminance output value).

The IDEX function has been applied to higher luminance values but has not been validated at those levels.

## 10.3 Display Maintenance

Unfortunately, for a variety of reasons, displays do not remain in calibration once they are set up. Any change in control settings (brightness and contrast) will, of course, change the monitor calibration and affect any perceptual linearization function that has been applied. If at all possible, contrast and brightness controls should be locked once calibration has been performed. Any significant movement of a CRT can affect its calibration, as can movement of any nearby equipment affecting magnetic fields.

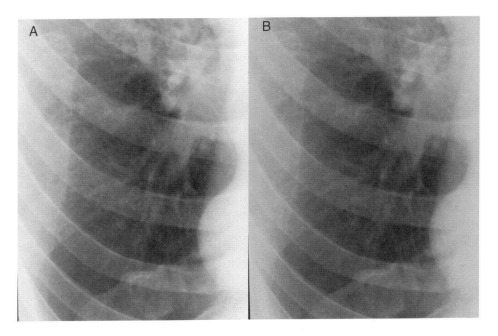

**Figure 10.12** Effect of decrease in Lmax.

Displays have a tendency to get dirty, and they should be cleaned regularly. Fingerprints and dirt reduce the effective contrast of the monitor. An image can be "burned into" a CRT if it is left on for a long period, so a screen saver or power saver option should always be used. If a power saver mode is used, the effect on warm-up time should be determined and the display should not be used for critical viewing during the warm-up period.

Displays age, so periodically they must be recalibrated. Generally, Lmax decreases over time. Figure 10.12 shows the effect of a decrease in Lmax. Image B shows the equivalent of a 34 $cd/m^2$ decrease in Lmax relative to image A.

Different types of displays degrade at different rates. There is no good rule of thumb regarding frequency of recalibration. Factors affecting frequency include the type and design of the monitor, operating conditions, criticality of calibration, and monitor age. More frequent calibration is required for CRTs that do not use a dispenser cathode technology for displays operating at high luminance levels, for display uses where contrast detection is particularly important, and for display equipment nearing the end of its life (as noted in Chapter 7, once DR falls below 22 dB, a monitor should be replaced). As an alternative to instrument measurement, the Briggs target can be used to assess, on a relative basis, a degradation of monitor DR. The target should be displayed and read on at least a weekly basis by the same observer as a means of detecting possible loss. Once a loss is detected (a decrease of 5 in the Briggs score), recalibration should be performed.

## 10.4 Summary

The choice of a monitor type depends on viewing tasks and applications, the viewing environment, and budget considerations. Once a monitor type (or range of types) has been selected, DR and Cm performance are probably the two most important attributes. A preliminary assessment can be made with test images such as the Briggs target and SMPTE target. Bit depth and SNR performance can also be important. The accompanying graphics card or controller used to drive the display must be compatible with both the display and the CPU. Clock frequency and the presence of a calibration capability are also important considerations.

The performance of a display system is strongly affected by the operating environment, particularly the surrounding light sources. For maximum performance, displays must be operated at low surrounding light levels. Otherwise, significant contrast losses, particularly at low luminance levels, will occur. Displays must be calibrated to a defined luminance range and a perceptual linearization LUT applied to optimize performance. Monitors degrade over time, so periodic recalibration (and ultimately replacement) are necessary.

## References

[1] J. C. Leachtenauer, A. Biache, and G. Garney, *Comparison of AMLCD and CRT monitors for Imagery Display,"* Technology Assessment Office, National Imagery and Mapping Agency, Reston, VA (1998).

[2] Eastman Kodak Co., *Quality Assessment Report for Imaging Displays,* Vol. 1(2), 1998.

[3] R. Myers, "Do digital interfaces make sense for CRT monitors?" *Information Display,* Vol. 7, pp. 14–17 (2001).

[4] Society for Information Display, "Annual directory of the display industry," *Information Display,* Vol. 16(8), pp. 37–83 (2002).

[5] Photonics Spectra, *The Photonics Buyers' Guide,* Book 2, 48th International Edition, Laurin Publishing, Pittsfield, MA (2002).

[6] M. H. Brill, "Color reversal at a glance," *Information Display,* Vol. 6, pp. 36–37 (2000).

[7] Society of Motion Picture and Television Engineers (SMPTE), *Specifications for medical diagnostic imaging test pattern for television monitors and hardcopy recording cameras,* SMPTE recommended Practice 133-1986 (1986).

[8] J. C. Leachtenauer, A. Biache, and G. Garney, "Effects of ambient lighting and monitor calibration on softcopy image interpretability," Final Program and Proceedings, PICS Conference, Society for Imaging Science and Technology, Savannah, GA, April 25–28, 1999, pp. 179–183.

[9] E. Muka, H. Blume, and S. Daly, "Display of medical images on CRT softcopy displays," *Proc. SPIE,* Vol. 2431, pp. 341–358 (1995).

[10] F. M. Behlen, B. M. Hemminger, and S. C. Horii, "Displays," in Y. Kim and S. Horii (eds.), *Handbook of Medical Imaging, Vol. 3, Display and PACS*, SPIE Press, Bellingham, WA, pp. 405–440 (2000).

[11] D. C. Rogers, R. E. Johnston, and S. M. Pizer, "Effect of ambient light on electronically displayed medical images as measured by luminance-discrimination thresholds," *J. Opt. Soc. Am. A*, Vol. 4(5), pp. 976–983 (1987).

[12] R. J. Farrell and J. M. Booth, *Design Handbook for Imagery Interpretation Equipment*, Boeing Aerospace Co., Seattle, WA (1984).

[13] National Electrical Manufacturers Association, *Digital Imaging and Communications in Medicine (DICOM) Part 14: Grayscale Standard Display Function*, Rosslyn, VA (2001).

[14] E. A. Krupinski and H. Roehrig, "Influence of monitor luminance and tone scale on observers' search and dwell patterns," *Proc. SPIE*, Vol. 3663, pp. 151–156 (1999).

[15] J. C. Leachtenauer and N. Salvaggio, "Color monitor calibration for display of multispectral imagery," Society for Information Display, International Symposium, Digest of Technical Papers, Boston, MA, May 13–15, 1997, pp. 1037–1040.

[16] C. Crites et al., *Visual Tonal Discrimination Characteristics and Their Effect on Tonal Display Calibration*, General Electric Co., Valley Forge, PA (1986).

[17] J. G. Rogers and W. L. Carel, *Development of Design Criteria for Sensor Displays*, Hughes Aircraft Company, Culver City, CA (1973).

[18] J. C. Leachtenauer, G. Garney, and A. Biache, "Effect of monitor calibration on imagery interpretability," Final Program and Proceedings, PICS Conference, Society for Imaging Science and Technology, Portland, OR, Mar. 26–29, 2000, pp. 124–129.

[19] P. J. G. Barten, *Contrast Sensitivity of the Human Eye and Its Effect on Image Quality*, SPIE Press, Bellingham, WA (1999).

[20] J. Montanaro, *Procedure for Calculating Monitor Compensation Look-up Table (LUT) from Monitor Calibration Measurements*, Veridian Corporation, Reston, VA (2000).

# Chapter 11
# Pixel Processing

One of the inherent advantages of soft copy viewing is the ability to manipulate an image in the spatial and tonal domains. With a hard copy, DR is restricted to the density range of the film or paper used. The tone (or color) displayed on the image is proportional to the light energy distribution in the scene and the characteristics of the capture and display devices (e.g., exposure, gamma, DR). In the soft-copy environment, the range of tones or colors can be manipulated in real time to enhance specific portions of the DR. Further, a variety of enhancement operations in the spatial domain can be applied. This chapter reviews and demonstrates the effects of various pixel processing operations by defining the order in which they should be applied for maximum effectiveness. Three categories of pixel processing are described: spatial filters, pixel intensity transforms, and geometric transforms. Bandwidth compression is addressed briefly so that the reader may become familiar with the concept, but the mathematics are not explained.

Pixel processing operations have also been developed to perform image reconstruction, segmentation, and feature or target identification. These operations are beyond the scope of this book. A review of these operations in the medical field is provided by Ref. [1] and a discussion relative to the surveillance and reconnaissance community in Ref. [2].

## 11.1 Pixel Intensity Transforms

Pixel intensity transforms involve the assignment of new intensity (CL) values to a pixel based on the original value of the pixel. With monochrome imagery, the resultant luminance value of the pixel is modified. With a color image/display, the colorimetric properties of the pixel may also be changed. This includes so-called "pseudocolor transform," where monochrome intensity values are assigned a color value.

### 11.1.1 Dynamic range adjustment

With monochrome imagery, two types of tonal adjustments can be applied. The first is called dynamic range adjustment (DRA). This typically entails expanding the range of the original image values to fit the full display DR. Figure 11.1 illustrates an example, including histograms, showing an image before and after DRA. The original image did not fill all of the available pixel values so contrast was reduced. The DRA process expanded the original values (78–255) to the range of

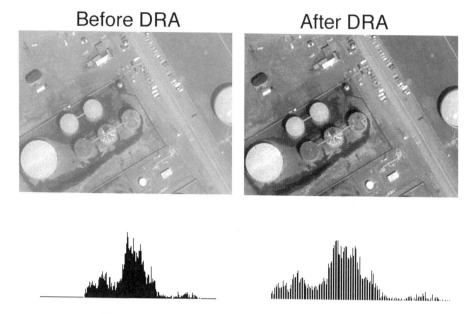

**Figure 11.1** An image shown before and after DRA.

0–255. The same number of unique pixel values was maintained, but they were spread out over the full display DR.

In this example, the operation was linear and was of the form

$$I_{new} = M(I_{old} - I_{old\,min}), \qquad (11.1)$$

where $I_{new}$ is the transformed value, $M$ is a multiplication factor, $I_{old}$ is the original value, and $I_{old\,min}$ is the minimum old value. In some cases, a constant based on the histogram may be used instead of the minimum value. The same process can be used when the image range is too dark. Note that this process should be used prior to any sharpening algorithm, since the sharpening algorithm will change the properties of the image histogram.

With some images, intensities above or below a certain value may be of no interest. A DRA can be used to clip such images such that all values above the region of interest are assigned the same intensity value. The term "histogram penetration" is sometimes used to describe the process whereby a defined percentage of pixel values at the top or bottom end of the histogram are clipped.

### 11.1.2 Tonal transfer adjustment/correction

The second type of pixel intensity transform performed on monochrome imagery is called tonal transfer adjustment (TTA) or tonal transfer correction (TTC). A TTA entails a linear or nonlinear transform of the original pixel intensity values as a

PIXEL PROCESSING                                                                221

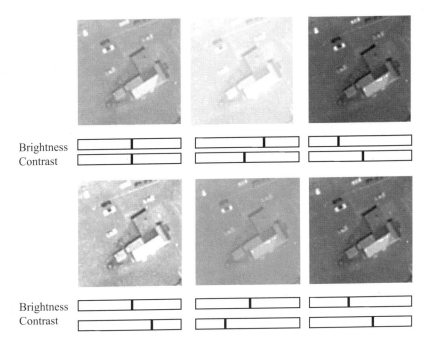

**Figure 11.2** Linear TTA.

means of expanding or emphasizing some portion of the DR. In the linear domain, brightness and contrast can be adjusted. Figure 11.2 shows the effect. The bars below each image indicate the position of a contrast and brightness adjustment slider bar. The original is shown in the upper left of the figure. The other two images in the upper row show an increase and decrease in brightness while contrast is unchanged. The two images to the left of the lower row show an increase and decrease in contrast while brightness is constant. The last image shows a decrease in brightness and an increase in contrast.

The effect of linear TTA operations can be described in terms of the image histogram. A change in brightness moves the histogram up or down the CL scale. Figure 11.3 shows the effect. The original image is shown in the upper left of Fig. 11.2. As brightness increases, an increasing number of pixels saturate at the maximum value, but the general shape of the histogram remains the same. Increasing contrast spreads the histogram. If the spread is sufficient, saturation also occurs. See also Fig. 3.17 for the effect of contrast and brightness on the display I/O function.

Nonlinear TTA provides more flexibility because specific portions of the DR can be expanded. Figure 11.4 shows an example. In image A, the low end of the DR has been compressed and the mid and high ends expanded. In image B, the low- and mid-tones have been expanded and the bright tones compressed. In image C, both the low- and high-end tones have been compressed and the mid-tones expanded. Some software packages provide families of nonlinear TTA operations, while others allow users to graphically design their own by selecting and moving one or more inflection points on an I/O function.

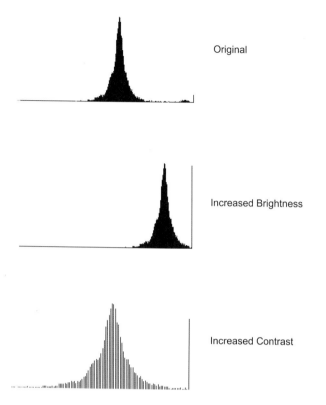

**Figure 11.3** Effect of linear TTA on an image histogram.

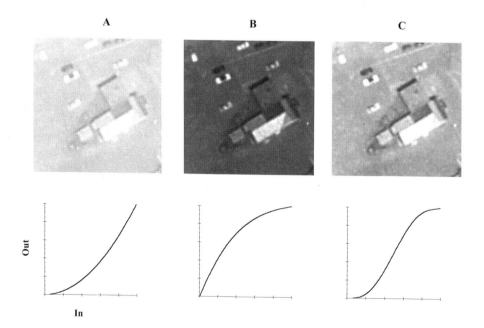

**Figure 11.4** Effect of nonlinear TTA.

PIXEL PROCESSING 223

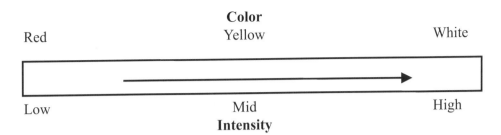

**Figure 11.5** "Heat" pseudocolor scale.

*11.1.3 Color transforms*

Color transforms can be applied to both monochrome and color images. The transforms, applied to monochrome images, substitute a color value for an intensity value. Because the HVS can potentially discriminate a large number of colors, it has been hypothesized that transforming monochrome to color images could improve tonal discrimination performance. The technique is generally known as pseudocolor, and a variety of schemes have been used. One of the more common is to treat the intensity scale as a flame scale as shown in Fig. 11.5. So-called random walk schemes have also been proposed where the color difference between adjacent intensities or tones is maximized.

The difficulty with pseudocolor schemes is that they may be confusing to the observer, thereby obscuring rather than enhancing the information content. This is particularly true for the nonsystematic pseudocolor schemes. In the reconnaissance and surveillance field, pseudocolor has generally been discarded as a useful technique. There are cases, however, where pseudocolor can be useful. For example, where intensity slicing can be used to discriminate objects of interest, pseudocolor be can used to enhance detectability. Also, when images from two imaging modalities are combined, color may be useful for indicating the source of the information in the fused image. In the medical field, color has been successfully used in this manner in Doppler ultrasound.[3] On the other hand, in a study where MRI and PET images were merged with different color scales, signal detection performance was degraded.[4] Another study showed that signal detection on gray-scale images was better than that on 12 different color scales.[5]

With color imagery, all of the DRA and TTA transforms previously discussed can be applied to the image as a whole. In addition, the contribution of the individual color channels can be modified. Hue, saturation, and lightness can also be individually modified. These operations generally do not affect the quality or information content of an image, but they can be used to enhance an image for publication. A final class of operations is designed to maintain the same or similar colors when moving from one display device to another (e.g., monitor to printer). These will be briefly discussed in Chapter 12.

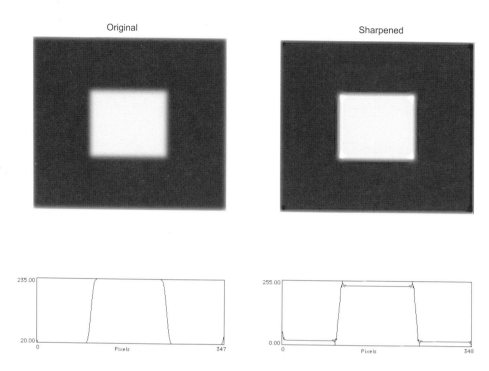

**Figure 11.6** Effect of image sharpening.

## 11.2 Spatial Filtering

Spatial filtering is used to sharpen edges in an image, to reduce noise, or some combination of both. In general, spatial filtering attempts to overcome deficiencies of the image acquisition system. For example, a high-frequency boost can be applied to overcome MTF losses in the acquisition system. Figure 11.6 shows the effect of sharpening an image. Intensity traces through the two images are shown along with the images. Spatial filtering should always be applied prior to any pixel intensity transform, since the spatial filter affects pixel intensities. Spatial filtering is typically applied by convolution in the spatial domain but can also be applied with multiplication in the frequency domain.[6]

Spatial filtering used to sharpen an image or compensate for MTF losses also increases noise. This is particularly apparent on synthetic aperture radar (SAR) imagery. Figure 11.7 shows a SAR image with and without MTF compensation (MTFC) applied.

Convolution is performed using a convolution kernel—an array of values such as that shown in Fig. 11.8. The kernel is placed over each pixel in an image, and the kernel value is multiplied times the image intensity value and then summed to represent the center value as shown in Fig. 11.8. Since the multiplication and addition operations increase the brightness of the pixel value, the kernel values are often scaled by the sum of their values (11 in the example shown). The example given in

PIXEL PROCESSING                                                                      225

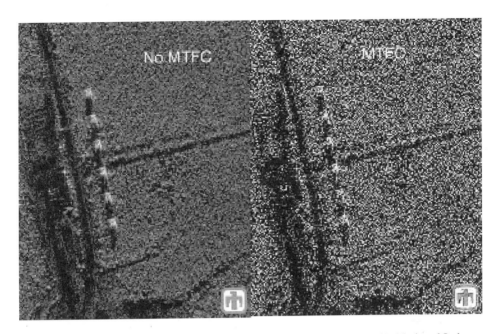

**Figure 11.7** Effect of MTFC on SAR imagery. Image courtesy of Sandia National Laboratories.

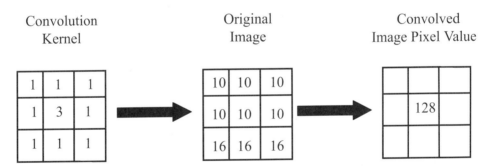

**Figure 11.8** Convolver operation.

Fig. 11.8 will have increased image brightness but blurred pixels because the values are averaged. Note that some technique is required to address the edges of the image since three image pixels are not present at an edge.

Now consider the array of image values shown in Fig. 11.9. The original array is shown on the left. The kernel in Fig. 11.8 is then applied, although, for the sake of this illustration, the values on the edge of the original are ignored. After applying the convolver, the values in the center result. These values are then divided by the sum of the values (11) in the convolution kernel to produce the values on the right. Because of the averaging that has taken place, the original image has been blurred and maximum values reduced.

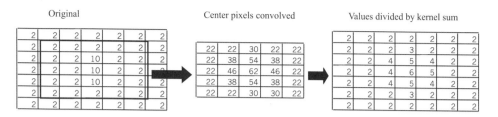

**Figure 11.9** Effect of convolution on image values.

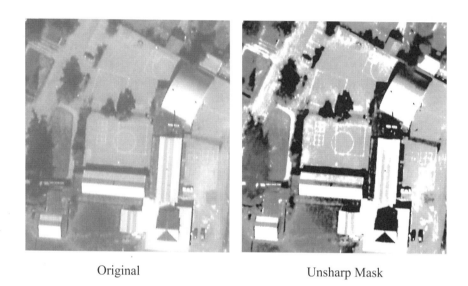

**Figure 11.10** Effect of unsharp masking.

The number of cells in the convolution kernel can be varied. When this number is increased, pixels a greater distance away from the center pixel are included. A kernel with the same value in each cell will blur the image; this is called a low-pass filter since it will tend to blur high spatial frequencies but not affect low frequencies. Kernels can be used to both sharpen and blur as a function of frequency. The so-called "unsharp masking" algorithm enhances high frequencies and leaves the low-frequency data unenhanced. It is accomplished by subtracting a low-pass frequency filtered image from the original image, multiplying by a gain factor, and adding the multiplied image to the original. Figure 11.10 is an extreme example.

Spatial filtering can also be performed in the frequency domain. Briefly, the process involves taking a two-dimensional FFT of the original image. The details of Fourier analysis will not be described here, but Fig. 11.11 shows an image and the magnitude of the FFT of that image. The FFT is rotationally symmetric and frequency increases away from the center of the FFT. We can multiply the FFT by a circularly symmetric filter that differentially multiplies each frequency or frequency band. Values less than 1 will reduce energy and values greater than 1 will increase

# PIXEL PROCESSING

**Figure 11.11** Right-hand image shows the FFT magnitude image of the left-hand image.

energy in the frequency band. For a more detailed discussion of frequency filtering, the reader is referred to Schott.[7]

## 11.3 Geometric Transforms

Geometric transforms place the image into a different plane or space without, in theory, changing the pixel intensity values. As will be seen, however, some transforms either eliminate data or require that data not in the original image be generated. A 2x magnification, for example, requires that each original pixel be shown as 4 pixels. Three new pixels must therefore be created.

Geometric transforms include magnification and minification, rotation, roam or translation, and warp. Magnification and minification (sometimes referred to as reduced resolution) increase or decrease the number of display pixels used to represent an original image pixel. Minification is used to increase the size of an image that can be displayed on a display of a given addressability. Minification typically deletes every $n^{th}$ pixel in two dimensions to produce a smaller image. Minification by factors of 2 typically uses simple pixel decimation or deletion; minification by other powers (e.g., 3) requires some type of interpolation. This approach works well so long as the imagery shows reasonably high autocorrelation values (where pixel intensity values change slowly) but may not work as well for image types where values are not as highly correlated. This is illustrated in Fig. 11.12. Note that several of the corner reflectors in the SAR image do not appear in the 0.5x version of the image.

With magnification (and often rotation and roam), new pixels that do not exist in the original image must be created. The new pixels may be integer or noninteger multiples of the original pixels. In its simplest form, an image is magnified by an integer amount (2x, 3x, etc.); each original pixel is represented by a number of pixels that is the square of the magnification ratio (e.g., 4 for 2x). Where the image is

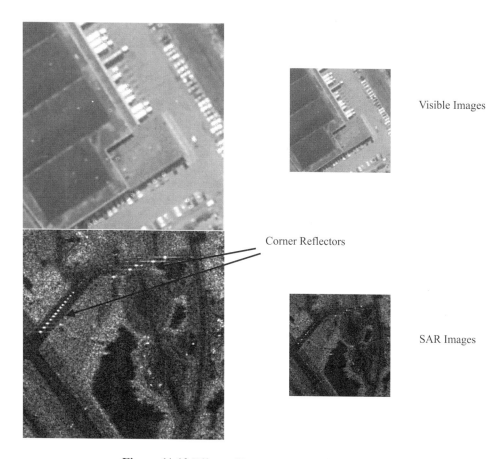

**Figure 11.12** Effect of image type on minification.

rotated or translated by less than a pixel, new pixels must be created that are a portion of two or more original pixels. In either case, some type of procedure is required to generate values for the new pixels. This process is called resampling.

The simplest form of resampling is pixel replication or nearest-neighbor resampling. With integer magnification, the pixel value of the original pixel is simply repeated to create the new pixels. A pixel that cannot be repeated by an integer pixel amount will adopt the value of the pixel closest to it in the original pixel space. Bilinear interpolation bases the new pixel value on the averaged values of the four surrounding pixels in the original image. It can thus be thought of as a 4 × 4 kernel centered on the new pixel. Bicubic interpolation applies a cubic polynomial in two dimensions to generate the value of the new pixel. The LaGrange interpolator uses a 1 × 4 kernel element in one dimension where the kernel values are based on the relative locations of the original and the new pixels.

Pixel replication and nearest-neighbor interpolation are the simplest and quickest methods of performing interpolation but result in a blocky-looking image if individual pixels cannot be resolved. Interpolations that average pixel values result in some blurring; the larger the size of the convolver kernel, the greater the time

# PIXEL PROCESSING

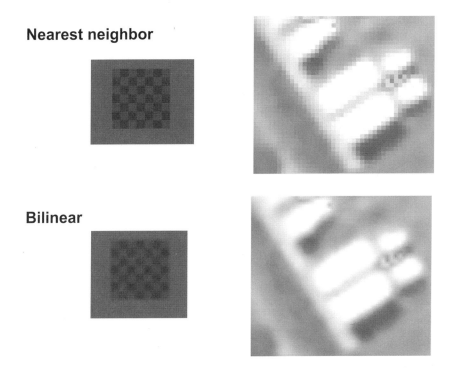

**Figure 11.13** Nearest-neighbor and bilinear interpolation.

required to perform the operation. Figure 11.13 compares a nearest-neighbor and a bilinear interpolation at 4x magnification for a Briggs target and an aerial image. Note the "blockiness" of the nearest-neighbor method and the blurring of the bilinear interpolation.

Where radiometric fidelity is important, pixel replication or even nearest-neighbor interpolation is preferred. These methods are also preferred where sharp edges must be preserved (as in the Briggs target).The LaGrange interpolator is preferred for most image applications where time (or computing power) is not an issue. In roaming operations, very little time is available to perform interpolation if the roam is to be accomplished at a reasonably fast pace, so bilinear or nearest-neighbor interpolation may be preferred. Where the level of magnification is 2x and addressability is high, the blockiness normally resulting from nearest-neighbor interpolation may not be evident. In a similar sense, it may be preferable for some applications to lower display addressability rather than require image magnification.

## 11.4 Bandwidth Compression and Expansion

The term "bandwidth compression" is a misnomer in that bandwidth is not compressed; rather, images are compressed to fit an available bandwidth. Bandwidth defines the amount of information that can be transmitted over a communication link in a defined period of time, usually expressed as bytes (8 bits) per second. Computer telephone modems can transmit data at 56 kilobytes/second (56,000 bytes per second). An image with 2000 × 2000 pixels and 256 gray levels (8 bits) would require 71 seconds to transmit. By compressing the image, transmission time can be significantly reduced.

Image compression capitalizes on redundancy in an image as well as the relative importance of low- and high-frequency information. For example, delta pulse code modulation (DPCM) encodes pixel intensity differences rather than the actual pixel values. Further, difference levels can be quantized (assigned one of a small number of values depending on the values of the differences). Visually lossless compression can be achieved with roughly a 50% reduction in image data. In one study, lossless compression rates for various medical imaging modalities ranged from 2.2 to 2.6.[1]

Operations in the frequency domain are more efficient. The JPEG discrete cosine transform (DCT) and JPEG 2000 algorithm (a wavelet compression) achieve higher compression rates. Lossless rates for medical imagery ranged from 3.4 to 3.5.[1] Figure 11.14 shows the NIIRS loss associated with various compression rates for the JPEG 2000 (wavelet) and JPEG DCT.[8] If you recall that a 1-NIIRS loss is roughly equivalent to a doubling of ground-sampled distance (GSD), the significant advantage of the JPEG 2000 algorithm over the JPEG DCT is apparent.

The impact of a given compression rate depends on the nature of the input imagery and the application. Figure 11.15 shows the effect of a 12x JPEG DCT

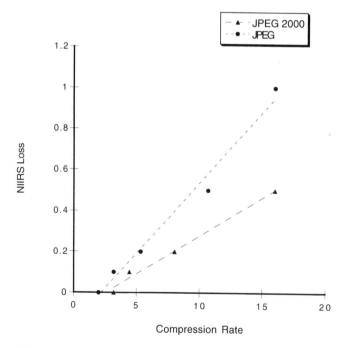

**Figure 11.14** Effect of compression algorithm and rate on NIIRS loss. Data from Ref. [8].

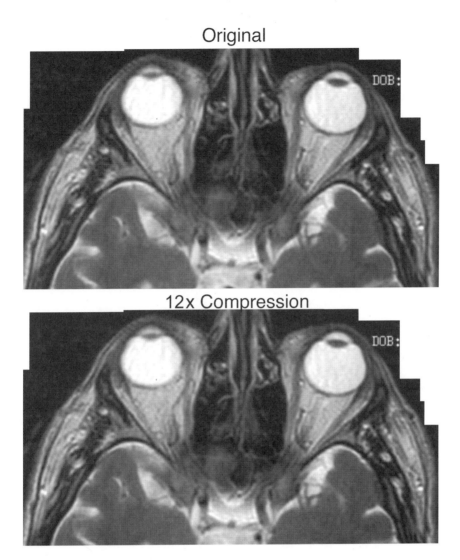

**Figure 11.15** Effect of JPEG DCT compression on MRI image.

compression on a portion of an MRI image. Differences are not obvious, although they are present and become more apparent in a difference image.

The effect of compression on chest images was studied at rates of 15:1 to 61:1. Although subjective quality was shown to decline, there was no statistically significant reduction in the ability to detect abnormalities.[9] A second study showed no significant difference in computed radiography (CR) images of the hand based on ratings of the depiction of hand features such as bone cortex and lesion margins.[10] A compressed image at a 20:1 rate was compared to an uncompressed image.

The examples shown thus far have dealt with monochrome images. Multichannel imagery (imagery collected simultaneously or sequentially at more than one wavelength band) such as multispectral and hyperspectral offer additional

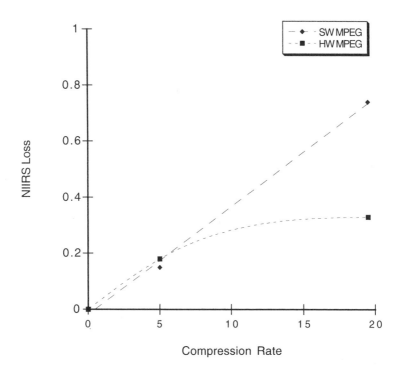

**Figure 11.16** Effect of MPEG-2 rate and implementation on NIIRS loss.

opportunities because of the redundancy across channels. Thus, rather than recording and transmitting 8 or more bits per channel, it is possible to record channel differences at a much reduced rate. In the case of motion imagery (e.g., video), compression can occur related to the temporal domain. As one moves from one frame to the next in a video sequence of ~30 frames per second, there is relatively little change in the scene. One can thus predict the contents of one frame from the previous frame, as opposed to storing the whole frame. The MPEG compressions (MPEG-1, MPEG-2, MPEG-4) employ this and other DCT-related approaches to compress video imagery. The MPEG-1 was designed to compress video to a CD-ROM; MPEG-2 and MPEG-4 are more robust in terms of rates and applications. Figure 11.16 shows the effect of MPEG-2 compression at various rates on NIIRS loss. The figure also shows performance for both a hardware (HW) and software (SW) implementation. The HW implementation has clearly been better optimized. Compare the data in this figure to those in Fig. 11.14.

Image compression requires an understanding of the statistical behavior of the image as well as an understanding of how the image is displayed and perceived. Quantizing image differences is most successful when the statistical distribution of those differences and their perceptual effects are known. It is also the case that rather large performance differences may be advertised without much substance. One should beware of claims of lossless compression at rates of 50:1 and greater. It is possible to disguise loss by the manner in which images are displayed and compared.

It is also necessary to be aware of unannounced compressions. Some software packages, particularly word processing and presentation, compress image data by factors of 2:1 or greater. In some cases, the compression is done with rather severe quantizing. For critical applications, users should compare input and output file sizes as well as histograms on the original and embedded images.

## 11.5 Sequence of Operations

Pixel processing operations should be performed in a defined sequence (Fig. 11.17).[12] If an image has been compressed, it must first be expanded. In some cases, the bit depth of the original image exceeds the bit depth of the processing algorithms. In this case, the data must be linearly compressed (usually) to the correct bit depth. For some systems, it is expedient to store a set of minified (reduced resolution) images. The next step in the processing chain is to apply the necessary interpolator and minify the images. If DRA is to be performed on an image based on its histogram, a histogram should be generated from the full-resolution image before additional processing is performed.

The next step is to perform the interpolations and convolutions needed to magnify, rotate, or sharpen the image. This step is performed before any pixel intensity transforms are made because these operations change the pixel intensity values. The next step is DRA, followed by TTA operations. Then the perceptual linearization LUT described in Chapter 10 is applied. If processing is performed at a bit depth greater than the capability of the final display, a linear compression of the image's values is the final processing step.

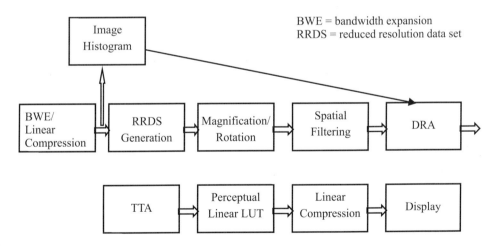

**Figure 11.17** Image processing sequence of operations.

## 11.6 Summary

Pixel processing operations entail spatial processing, intensity changes, and geometric operations. Bandwidth compression and expansion are also forms of pixel processing. Spatial filters are used to sharpen images or to reduce noise and artifacts. Spatial filtering is performed by convolution in the spatial domain or multiplication in the frequency domain. Pixel intensity transforms are accomplished to spread image data over the total dynamic range (DRA) and to differentially emphasize portions of the intensity scale (TTA). With color imagery, additional transforms can be applied to individual color channels to change the color space or the appearance in color space.

Geometric transforms are applied to perform operations such as magnification, rotation, and roam. These processes require that "new" pixels be generated to fill in gaps in the pixel matrix. Some form of interpolation based on values of surrounding pixels is used to perform these operations. Finally, bandwidth compression is the term applied to image compression operations that are performed to fit data to an available bandwidth. Bandwidth compression uses a variety of techniques that capitalize on within-scene image redundancy. Compression of multiband and video imagery also capitalizes on spatial, tonal, and temporal redundancy.

The sequence in which pixel processing operations are performed has an important effect on image quality. After bandwidth expansion (if compression has been performed), geometric transforms are first accomplished. Spatial filtering is performed next, followed by pixel intensity transforms (DRA followed by TTA).

## References

[1] P.W. Jones and M. Rabbani, "JPEG compression in medical imaging," in Y. Kim and S. Horii (eds.), *Handbook of Medical Imaging, Volume 3, Display and PACS*, SPIE Press, Bellingham, WA, pp. 221–276 (2000).

[2] J. C. Leachtenauer and R. G. Driggers, *Surveillance and Reconnaissance Imaging Systems: Modeling and Performance Prediction*, Artech House, Boston, MA (2001).

[3] E. A. Krupinski, "Practical applications of perceptual research," in J. Beutel, H. L. Kundel, and R. L. Van Metter (eds.), *Handbook of Medical Imaging, Volume 1, Physics and Psychophysics*, SPIE Press, Bellingham, WA, pp. 895–929 (2000).

[4] K. Rehm et al., " Display of merged multimodality brain images using interleaved pixels with independent color scales," *J. Nuc. Med.*, Vol. 35, pp. 1815–1821 (1994).

[5] H. Li and A. E. Burgess, "Evaluation of signal detection performance with pseudocolor display and lumpy backgrounds," *Proc. SPIE*, Vol. 3036, pp. 143–149 (1997).

[6] Central Imagery Office, *USIS Standards and Guidelines*, CIO-2008, Vienna, VA (1995).

[7] J. R. Schott, *Remote Sensing: The Image Chain Approach*, Oxford University Press, New York (1997).

[8] S. D. Rajan, A. T. Chien, and B. V. Brower, "Advanced commercial compression that will enable USIGS to meet the requirements of TPED and JV2010," paper presented at the 1999 IS&R Conference, Washington, DC (1999).

[9] W. F. Good et al., "Observer performance of JPEG-compressed high-resolution chest images," *Proc. SPIE*, Vol. 3663, pp. 8–13 (1999).

[10] K. Uchida et al., "Clinical evaluation of irreversible data compression for computed radiography of the hand," *J. Dig. Imaging*, Vol. 11, pp. 121–125 (1998).

[11] J. C. Leachtenauer, M. Richardson, and P. Garvin, "Video data compression using MPEG-2 and frame decimation," *Proc. SPIE*, Vol. 3716, pp. 42–52 (1999).

[12] Eastman Kodak Company, *Image Chain Analysis Softcopy Image Processing Chain Baseline*, 1998.

# Chapter 12
# Digitizers, Printers, and Projectors

Previous chapters covered the process of image display in soft copy. This chapter covers primarily the process of hard-copy input and output. Although many imaging modalities have become purely digital with direct input to the image workstation, film still remains a common recording medium. The film must be digitized using some type of scanner or digitizer in order to be displayed on a soft-copy workstation. The term "digitizer" refers to any device that is used to convert a hard-copy transparency or reflective print to a string or array of digital values. The term "scanner" implies a digitizer where either the hard-copy input or detector is moved (scanned), but the term is often applied to the generic process of digitizing.

Although digital projection displays are in common use, the need still arises to print digital output data to a hard-copy medium, either paper or transparency. Film also remains as an efficient storage medium.

This discussion begins with device operation for hard-copy input and output. This is followed by a discussion of image quality as it relates to device selection. Finally, guidance is provided on how best to use hard-copy input and output devices.

Because digital projection displays are becoming a common means of presenting digital imagery to large groups, projection display quality parameters and image processing for those displays are discussed briefly.

## 12.1 Digitizers

Digitizing devices convert transmitted or reflected light energy to some form of electrical energy. The electrical energy is then quantized (in intensity and sometimes space, depending on the capture method) to form a digital bit stream. The bit stream is assembled in the same geometry as it was captured to form a digital image.

### 12.1.1 Digitizer operation

Devices used to convert hard copy to a digital bit stream are of four types:

1. Flatbed and drum microdensitometers;

2. CCD frame grabbers or array scanners;

3. CCD flatbed linear scanners; and

4. Laser digitizers.[1]

A microdensitometer uses a photomultiplier (PM) tube to measure light energy. A PM amplifies detected light energy and converts the energy to an electrical signal that is then digitized. The film to be digitized is illuminated and moved past the fixed tube. In a drum microdensitometer (scanner), the film is fastened to a drum that moves around and then translates past the fixed tube. With a flatbed scanner, the film is flat and translates in $x$ and $y$ past the tube. Optics are used to define the resolution of the scanner.

A CCD frame grabber or array scanner operates in a manner virtually identical to a digital camera. A lens is used to focus an illuminated print or transparency onto a CCD array. The CCD converts the light energy to electrical energy that is then digitized. The digitized image is thus an array of pixels having the same number of pixels as in the CCD array. With some array scanners, either the film or the CCD array is translated so as to tile together a larger image than can be captured by the single CCD array.

A CCD flatbed linear scanner moves a CCD array across the illuminated input medium. The size of the array in one dimension defines the size of the captured image in that dimension. The length of the scan defines the other dimension.

A laser scanner moves a laser beam (a coherent narrow-band energy at a specific wavelength) across the film image and captures the transmitted laser energy. Three-dimensional laser digitizers also exist that use laser energy to map distance (and thus surface shape) and light energy to capture color information.[2]

Charge-coupled devices have a mostly linear response to light energy: the digitized image is linear with respect to film transmission but nonlinear with film density. Photomultipliers have a nonlinear response to light energy and a linear response to film density. A comparison of CCDs and PMs is shown in Fig. 12.1. The PM response is more or less linear with film density. The CCD response is nonlinear, which tends to compress the low- and high-density areas as shown in Fig. 12.2. Note the loss of detail in the light-toned areas. Images captured with a CCD will often require tonal adjustment. Laser digitizers may use a logarithmic amplifier to avoid compression at high densities.[2]

Digitizers can also be characterized in terms of the input medium type (film, reflective print, solid surface) and size. Flatbed scanners sold on the consumer market are designed primarily for paper print input, but some accept transparencies. Slide scanners are designed to accept a particular size of film. Digitizers for the medical market emphasize tonal accuracy as opposed to spatial resolution. Finally, digitizers can be characterized as color or gray scale.

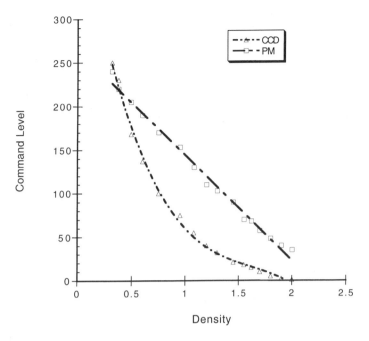

**Figure 12.1** Comparison of PM and CCD response.

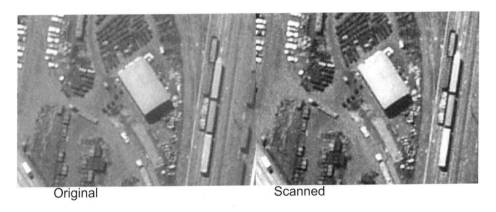

**Figure 12.2** Effect of nonlinear digitizer response.

### *12.1.2 Digitizer image quality and device selection*

The quality of digitizers can be characterized in terms of spatial fidelity, radiometric fidelity, and geometric fidelity.[3] Ideally, a digitizer should accurately capture all of the information that is in the original hard-copy image and portray that information with no random or systematic errors.

Spatial fidelity is used here as a measure of the degree to which all of the fine detail in the original image is captured in the digitized image. Digitizer resolution

Table 12.1 Resolution conversion.

| Dots/in. (dpi) | Spot size (μm) | Film resolution (cy/mm) |
|---|---|---|
| 6350 | 4 | 125 |
| 4233 | 6 | 83 |
| 3175 | 8 | 62 |
| 2117 | 12 | 42 |
| 1693 | 15 | 33 |
| 847 | 30 | 17 |
| 605 | 42 | 12 |

is used as the physical measure of performance. Digitizer resolution is typically defined in terms of dots per in. (dpi) or in micrometers (μm). Both measures relate to the size of the digitizer capture element. Film resolution is more often defined in terms of cycles per millimeter (cy/mm). Table 12.1 shows conversion values. Film resolution may also be defined in terms of a writing spot size if the hard copy has been generated from a digital input using a digital writing device. In that case, the column labeled "μm" would correspond to the writing spot size.

Digitizer resolution is sometimes defined as both optical and interpolated resolution. Optical resolution is based on the projected size of the CCD element or PM tube; interpolated resolution is simply a digital enlargement (with interpolation) of the scanned image. Interpolated resolution figures should be disregarded. Flatbed scanners typically provide optical resolution values in two dimensions, e. g., 1200 × 2400 dpi. The first number refers to detector (horizontal) spacing, the second to the number of scans per in. (vertical spacing). There is thus overlap in the vertical dimension. This is not the case with frame grabbers.

Although in theory digitizing should be performed at two samples per input resolution element, little or no information loss may occur with coarser sampling. Figure 12.3 shows the NIIRS loss associated with various digitizing spot sizes.[3] The original film was at a resolution of 120 cy/mm. At two samples per cycle, the nominally required digitizing spot size would be 4.2 μm. A sample of 25 images was scanned on eight different digitizers at various resolution settings. The digitized images were displayed and rated on a high-quality monochrome CRT. Again, in theory, a doubling of resolution should result in a NIIRS loss of 1. As shown in Fig. 12.3, this was clearly not the case. Although there was a strong linear relationship, the slope was less than expected. Increasing the digitizing spot to 8 μm resulted in less than a 0.2 NIIRS loss. Roughly a seven-fold increase was required to produce a NIIRS loss of 1.

Again referring to Fig. 12.3, multiple data points at the same spot size came from different digitizers. Performance was thus largely due to spot size as opposed to digitizer model. The range across models was 0.2 NIIRS or less.

Radiometric fidelity is characterized using four measures—DR, linearity, large-scale uniformity, and noise (or SNR). Dynamic range refers to the ability of the digitizer to capture the full DR of the input media. Film DR is measured in

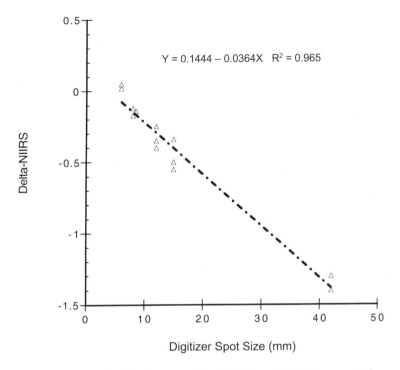

**Figure 12.3** Relationship between delta-NIIRS and digitizing resolution.

terms of the density range of the film. Medical imagery often has a wide range with maximum film densities as high as 4.0 and minimums as low as 0.2 density units. Note that a density of 4.0 is equal to a transmittance of 0.01%. Aerial imagery typically has a lower maximum density, on the order of 2.5 or less. If the digitizer does not have a sufficient DR, low densities (dark tones) will not be separated. The nonlinearity of CCDs was previously discussed. Although an LUT can be applied to the digitized image, information will still be lost.

Large-scale uniformity refers to intensity variations across the format of the digitized image. Because digitizer light sources vary as a function of position, it is not uncommon for large-format digitizers to show intensity nonuniformity.

Noise in the digitizing operation can result from electronic noise as well as dirt on the media or digitizer surfaces. Figure 12.4 shows noise measurements made on three different digitizers.[1] A step wedge was digitized and the resultant image scanned. Intensity differences across the scan were converted back to density units using the known digitizer transfer function. The variability in density units provides an indication of the number of discriminable density values. In the example shown, scanner #2 would be expected to have half the number of discriminable density steps as scanner #1.

Although not a direct measure of radiometric fidelity, bit depth can be expected to show a correlation with fidelity. This is particularly true if radiometric correction is required. Scanner bit depth typically ranges from 12 to 16 bits per channel.

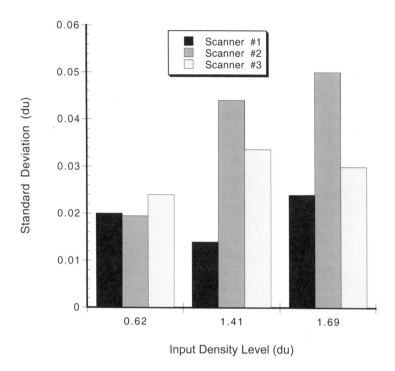

**Figure 12.4** Digitizer noise measurements.

Geometric fidelity refers to the ability of the digitizer to accurately capture position or location. For any device where the film or detector is moved, the possibility exists for nonlinear movement. In practice, however, this has not been seen as a problem, perhaps partly due to the inherent nonlinearities of the film/camera system. Geometric correction to internal or external control points is usually necessary; this can be done with the digitized image even if the digitizing process introduces distortions. Such distortions are not, in the author's experience, visible to the eye.

Two additional considerations affect the selection of a digitizer: the size of the images to be digitized and the volume of digitizing to be performed.[4] Film digitizers generally limit the physical size of the images that can be scanned. Frame grabbers limit size in terms of the number of pixels in the CCD array. Frame grabbers can have arrays as large as 4K × 4K. A 4-μm digitizing spot size limits the image capture area to a 0.62-in. square. A tiling capability is required to capture larger areas. In addition, the size of the image, coupled with digitizing resolution, defines the size of the image file. The 4K × 4K frame grabber at 12 bits per pixel produces a 24-MB file. A single 35-mm film frame digitized at 4-μm resolution produces a 90-MB file. Seven such images would fill a CD ROM. Images up to 8.5 × 11 in. can be scanned at resolutions on the order of 1600 × 3200 dpi using consumer-quality flatbed scanners. Special-purpose scanners can operate at resolutions as small as 4 μm (6350 dpi).

As might be expected, there is a resolution/cost trade-off. Digitizers range in price from a few hundred dollars to several thousands of dollars (U.S.). Flatbed scanners with resolutions on the order of 1200 × 2400 dpi can be purchased for a few hundred dollars. Film scanners with 4000-dpi resolution cost on the order of $1000 to $3000. Drum scanners cost $7000 and up.

The volume of digitizing produces two effects. First, the actual time required to scan an image is only a portion of the total digitizing time because the film must first be mounted on or in the digitizer, the digitizer must be initialized, and a pre-scan must be made to check the settings. Second, after the scan is performed, the data must be written to storage. In a previous study, the total time required to scan 5-in. film flats with a 4-µm spot ranged from 26 to 48 minutes. For large volumes of data, any features that reduce load, pre-scan, and write times can be advantageous.

### *12.1.3 Digitizing procedures*

The process of digitizing an image can be described in terms of four operations: film annotation and loading; initialization and prescan; scanning; and storage and verification. Although these operations are described in terms of film scanning, they are similar for paper prints and drawings. It is often the case that the area to be digitized is less than that of the total image, so if the digitizer operator and scan requestor are not the same person, a means will be needed to specify the area of interest to the operator. Although this seems obvious, accurately communicating the location of the 1/4- to 1/2-in. area to be digitized is not a trivial task. Rather than using transparent overlays or marking the film itself, the scan requestor should provide a low-resolution annotated scan. For example, the total image can be scanned by the requestor at 300 dpi to create a low-quality annotated print for the operator. The digitizer will normally be connected to a workstation with a viewer, so the operator can then adjust the area to be scanned.

Both the film and the digitizer surfaces should be cleaned before digitizing. Film should be loaded so that the emulsion side (the less shiny side) is mounted closest to the energy source. Ideally, film should be handled with gloves and stored (but not digitized) in a protective cover.

Once the film is loaded, the digitizer must be initialized. This process requires identification of the input media type, the desired digitizing resolution, and in some cases, specifications relating to the input density/output intensity function. Digitizing resolution should be based on the intended use of the final product. Table 12.2 provides guidelines for digitizing imagery.[5]

For near-lossless archival storage, the digitizing resolution should be set at twice the resolution of the original. For example, film with 120 line pairs per millimeter (lp/mm) resolution should have a digitizing spot size of about 4 µm or 6250 dpi. For paper prints with a resolution of 10 lp/mm, the spot size should be 42 µm (about 600 dpi). For halftone printing and viewing applications, the resolution should be on the order of twice the final display resolution, taking into account the

**Table 12.2** Digitizing resolution guidelines.

| Application | Recommended Resolution |
|---|---|
| Archival storage | Two samples per resolution element |
| Laser printer | 600–1200 dpi |
| Inkjet printer | 300 dpi |
| Email viewing | 150 dpi |
| Screen display | 100 dpi |

magnification of the original and the viewing conditions. Halftone printing consists of patterns of dots to represent images. Each dot is either "on" or "off"; as the number of "on" dots per area increases, the image becomes darker. The dots are small relative to the resolving power of the eye, thus we tend to see continuous tones as opposed to the actual dot patterns. For a given halftone, the number of dots per inch is fixed. As dot density increases, the apparent quality of the halftone image also generally increases. For example, with an 85-dpi halftone output, resolution should be at 170 dpi if the final is to be viewed at the same scale as the original. Digitizing resolution should also be at a value evenly divisible into the maximum optical resolution of the digitizer. Otherwise it will be necessary to interpolate to fill in the noninteger pixel values. For example, if the digitizer had a resolution of 600 × 1200 dpi and the desired output resolution was 170 dpi, a value of 200 × 400 would be used to ensure no resolution loss. A value of 150 × 300 could also be used at the risk of some loss of detail.

If the final copy is magnified relative to the original, the digitizing resolution needs to be increased by a factor equal to the enlargement factor. For example, assuming the final output of 170 dpi represents a 4x enlargement of the original, a digitizing resolution of 680 dpi should be used. With a 600 × 1200 digitizer, some loss of detail may be evident in one dimension.

The degree of magnification necessary depends on the intended use of the scanned image. If the image is to be viewed directly and the user assumes a 10-in. viewing distance, the resolution of the scanned image should be on the order of 260 dpi. This is equivalent to a resolution of 44 cy/deg and the commonly stated HVS resolution capability of 10 lp/mm for high-contrast detail. If we want to illustrate a particular feature in the image, it may be necessary to enlarge the feature. As a general rule of thumb, the feature should be subtended by at least 12 resolution lines (6 cycles) across the minimum dimension of the feature for a 50% probability of recognition.[6] Twice that value (24 lines or 12 cycles) is needed for a 97% probability. Figure 12.5 shows images digitized at four resolutions. Although the image at 12 lines may be identifiable, the image at 24 lines presents a clearer illustration of the tanks.

At 44 cy/deg, a 12-line feature will subtend about 8 minutes of visual arc. For low-contrast detail, a larger enlargement factor is required. From the J curve in Fig. 6.9, contrast sensitivity is greatest at about 2 cy/deg. This represents the extreme. From the data shown in Fig. 8.1, a screen resolution on the order of 75–

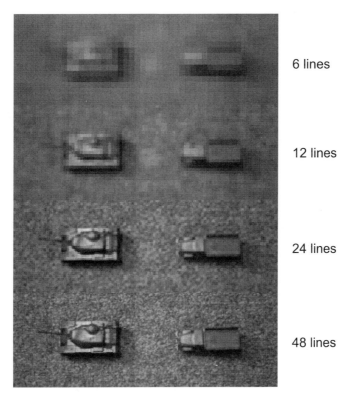

**Figure 12.5** Effect of subtending resolution lines.

80 ppi provided the best performance. This is equivalent to a value of roughly 11 cy/deg, or one-fourth the maximum. At 11 cy/deg, the 12 resolution lines should subtend 32 minutes of arc and the 24 lines about 1 degree. As a starting point to determine digitizing resolution for an enlarged image, assume a final resolution of ~260 dpi. Measure the minimum dimension of the object of greatest interest in the scene. That dimension should be at least 1/8th-in. on the final print and should subtend 30 minutes or more.

Figure 12.6 shows three examples where the vehicle and aircraft fuselages subtend the indicated visual angles at a 16-in. viewing distance. Note that the 45-minute example is easiest to see. The rough dimensions of the fuselage widths are 3/16th, 1/8th, and 1/16th-in., and the fuselage widths are covered by 18, 12, and 6 dpi, respectively. If we assume that the original fuselage width was 0.03 in., we need an enlargement factor of 4 (1/8th in.) to 6 (3/16th in.). Since the final resolution will be 260 dpi, the scanning resolution should be 4 to 6 times the final, or 1040 to 1560 dpi. Some adjustment may be required to maintain an integer scanning resolution factor. If we scan at 1040 dpi, each scan line is .0009 in. and we end up with 31 lines across the fuselage width—more than enough.

To summarize, the following steps must be performed to obtain the digitizing resolution:

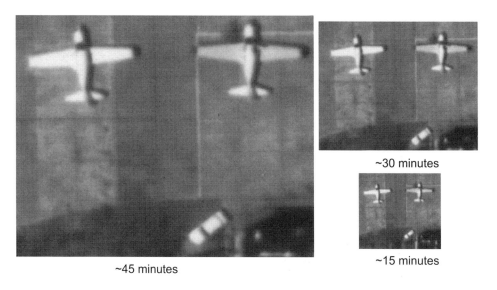

**Figure 12.6** Object subtense examples.

1. Define the desired output resolution in dpi;

2. Define the minimum dimension of the object of interest, $S_O$;

3. Define the desired size of the minimum dimension on the output image, $S_F$;

4. Define the magnification factor, $M$, by

$$M = \frac{S_F}{S_O}; \qquad (12.1)$$

5. Multiply the desired output resolution by the magnification factor, $M$, to obtain the digitizing resolution.

Another method of determining digitizing resolution requires a variable power magnifier and measuring device.[1] This method may be easier to use for some applications. Start by viewing the original and determining the size of the smallest detail in the image. Divide half this dimension (in inches) into 1 to define the required digitizing resolution in dpi. For example, if the smallest detail is 0.005 in., the required digitizing resolution is 400 dpi [1/(.005/2)]. Next, determine the magnification level needed to comfortably see the object(s) of interest (the required magnification) as well as the magnification needed to see the smallest detail in the image (the desired magnification). The desired magnification will always be equal to or greater than the required. Divide the higher number by the lower. This result defines the digitizing spot size enlargement factor. Divide this number into the previously defined digitizing resolution value to define the final approximate digitizing resolution. Remember to make the final value an integer multiple of the digitizer maximum resolution. As an example, assume that 4x magnification comfortably shows the object of interest but 12x is desired to see all of the image detail. The digitizing spot size enlargement factor is thus 3 (12/4). Dividing this into the previously established value of 400 dpi gives a value of 133 dpi. Assuming a digitizer with a resolution of 1200 × 600 dpi, use a final value of 150 dpi since 600 is an integer multiple of 150.

If the digitized image is to be projected, both the projection enlargement factor and the viewing distance of the audience must be considered. The relationship is shown in Fig. 12.7. If a viewgraph is to be used, it might normally be printed at a resolution of 300 to 600 dpi. The viewgraph would then be projected onto a wall or screen. A 10-in. wide viewgraph projected onto a 60-in. screen represents an enlargement factor of 6. From Fig. 12.6, our 1/8th-in. wide fuselage (30-minute subtense) is 0.75-in. wide (1/8th × 6). If the viewer is seated 10 ft from the screen, the 0.75-in. wide object subtends 21 minutes and thus appears smaller than if the viewgraph is viewed from a 16-in. distance. Therefore, we need a slightly larger enlargement factor than was used in Fig. 12.6. If the viewing distance is increased to 20 ft, the 0.75-in. dimension would subtend only 11 minutes, so we would need to double or triple the magnification ratio.

The viewgraph does not need to be printed at a high resolution if it is to be viewed from some distance. The previously cited requirement for 44 cy/deg at a 10-in. viewing distance applies to high-contrast detail. The image will appear degraded at lower resolutions. In a projection situation, viewing distance is generally fixed at some distance substantially greater than 10 in. For example, at a distance of 6 ft, we could theoretically write at 36 dpi (and consequently digitize at a 7x lower resolution). The problem with this approach is that the size of the viewgraph is fixed (at ~7 × 10 in.), and if we digitize at a very coarse resolution, the portion of

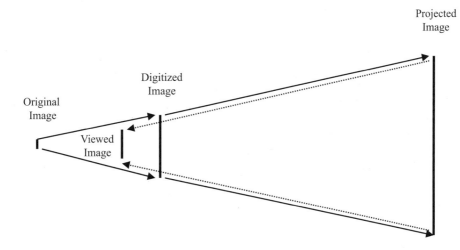

**Figure 12.7** Projection display relationships.

the original image that can be displayed is limited. We must strike a balance between object size (most important), scene context (often important), and digitizing and writing resolution (less important).

In the case of direct optical projection, resolution is limited by the native resolution of the projector. This is typically on the order of 1024 × 768. The projector uses an LCD array or CRT that may be optically magnified and is then projected onto a screen or wall. Here, the digitizing rules for screen display apply (100 dpi) with the necessary modification to cover magnification and viewing distance. If we assume a 6-ft wide screen and a 1024 × 768 projector, the projected display is at about 14 dpi. In order to maintain a viewing resolution of 44 cy/deg, we need to limit viewing distance to 3–4 ft.

Once the digitizing resolution is established, a number of possible tonal or color adjustments may be made. They include the same basic adjustments described in Chapter 11 and for monochrome imagery, including brightness and contrast adjustment (also sometimes called exposure and gamma on digitizers), DRA, and TTA. For color imagery, TTAs may be available separately for each color channel. Typically, these adjustments are made after a pre-scan has been performed. The pre-scan provides a thumbnail image that can be used to make adjustments. This method is relatively crude, so further adjustment may be needed using the image histogram as guidance. The goal in performing the pre-scan adjustments is to preserve as much of the data in the original as possible while recognizing that quantization losses will occur due to nonlinearities in the scanning process. Many digitizers have a bit depth greater than the final file or display bit depth. Less loss will occur if adjustments are made in the digitizer pre-scan process instead of on the final digitized image.

After the pre-scan and necessary adjustments have been completed, the scanning operation takes place. This requires anywhere from a few seconds to several minutes. The scanned image should be viewed in soft copy before it is saved to a file. It should normally be saved in uncompressed standard format (e.g., TIFF) or in a lossless or near-lossless JPEG format. The JPEG compression will trade decreases in image size for increases in artifacts.

## 12.2 Printers

Printers used to produce digital hard-copy images can be grouped into four categories based on the printing mechanism involved: silver halide, thermal, inkjet, and electrostatic (laser).[7,8]

In the discussion that follows, coverage is largely restricted to what are called secondary applications. The hard-copy output is not used for primary information extraction, but rather, to illustrate or demonstrate what has been observed in a soft-copy image. Further, the emphasis is on small work group or desktop applications where simplicity of operation is important.

## 12.2.1 Printer operation

Silver halide printers use a laser or light source to illuminate a silver halide-coated medium. The energy source is modulated in accordance with the digital image values and is scanned across the paper or film. Since silver halide printers require conventional photographic wet processing, they are seldom used for secondary hard-copy applications and will not be discussed further.

Thermal printers use an array of heating elements (controlled by the digital values of the image) to transfer a digital image to a heat-sensitive medium. The heat-sensitive medium then deposits dye on an image medium (print or transparency). Direct thermal printing is used for label printers but requires special paper, and has thus given way to inkjet and laser printers. A type of thermal printing technology called dye sublimation or thermal dye diffusion is used for high-end thermal image printers. Dye sublimation printers use three (C, M, Y) or four (C, M, Y, K) ribbons to transfer dye to the final image. The density of the dye is continuously variable over 8 bits per color, so thermal printers can match photographic film quality. They are relatively slow, however, since four passes are required, and the cost per page is relatively high.

As the name implies, inkjet printers deposit ink on the image medium. Ink is forced from tiny holes. Current inkjet printers use four colors (C, M, Y, K) and the ink is contained in two print heads (one for black and one for the three colors) that move on a rail across the page. The page then moves to the next line and the process is repeated. Current inkjet printers also approach or match photographic quality, although longevity of the finished product is an issue.

Finally, electrostatic printers operate by the discharging of a surface with a scanning laser beam. The beam is modulated by a controller in proportion to the digital image file to be printed. After the surface has been discharged, oppositely charged toner comes in contact with the surface and is attracted to areas retaining a charge. The toner is then fused to the surface or transferred to a second surface (the final output) and fused. Color laser printers use C, M, Y, K toners applied successively to produce a color image. Because color laser printers have, in essence, four times the complexity of a monochrome printer, they are considerably more expensive. However, the cost per page is quite low and the speed is relatively high. Current high-end laser printers also approach photographic film quality.

## 12.2.2 Printer quality and selection

The factors used to characterize printer quality are the same as those used for digitizers. Printer quality can be described in terms of spatial fidelity, radiometric fidelity, and geometric fidelity. Ideally, the printer should display the same image in hard copy as we see in soft copy, but this is seldom precisely the case. The goal is to approach this ideal as closely as possible.

Spatial fidelity is a measure of the degree to which all of the fine detail in the original digital image is displayed in the printed hard-copy image. Printer resolution is typically used as the physical measure of performance and is defined in terms of dpi. Whereas digitizer resolution translates to image pixels, printer resolution is more complex. Laser and inkjet printers are characterized in terms of dpi resolution. Laser printers vary dot density using halftoning to produce tonal variations. Inkjet printers vary dot density in a less regular fashion. Thermal printers vary exposure using a regular dot pattern. Figure 12.8 shows a portion of a Briggs target (original) and the same target written with a laser printer, an inkjet printer, and a thermal printer. The images have been enlarged to illustrate the differences. At normal viewing distance, the differences are much less apparent.

Printer resolution for inkjet and laser printers is also affected by the media on which the image is printed. With inkjet printers, the ink can expand as it hits the media and thus increase the spot size. Photographic-quality paper should be used with inkjet printers when printing images.

Current inkjet printers offer resolutions of up to 2400 × 1200 dpi and laser printers offer 1200-dpi nominal resolutions. In both cases, it takes several dots to form a pixel. In a practical sense, at a 10-in. viewing distance, the eye can see no more than ~500 dpi at high contrast. Imagery typically has lower contrast, and the

Original

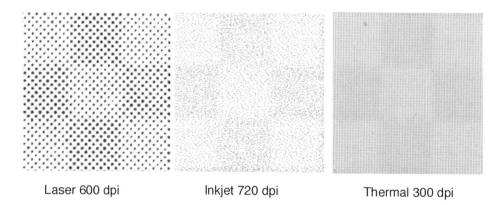

Laser 600 dpi                Inkjet 720 dpi                Thermal 300 dpi

**Figure 12.8** Effect of printer type on a portion of a Briggs target.

eye is less capable of seeing fine detail at lower contrast. Maximum contrast sensitivity is at 2 cy/deg. From the data in Fig. 8.1, 75 ppi at 16 in. is equivalent to 11 cy/deg at 10 in., which corresponds to a requirement of 126 dpi. This is approximately what can be produced on a 600-dpi laser printer (remembering that it takes multiple dpi to reproduce a pixel in halftone).

Radiometric fidelity is characterized using four measures: DR and gamut, tonal/color transfer, large-scale uniformity, and noise (or SNR). In this case, dynamic range refers to the ability of the printer to capture and display the full DR of the image. Output DR is measured in terms of the density range of the film or paper. Density is defined as

$$D = \log_{10} \frac{1}{T}, \quad (12.2)$$

where $T$ is transmittance or reflectance. Ideally, the printer output should preserve the intensity information in the original image. An 8-bit image has, in theory, a 24-dB DR. The actual number may be less due to nonlinearities in the process. On a calibrated display (0.34 to 120 cd/m$^2$), the DR is 25.4 dB. Paper and film media often have substantially lower DRs. The DR of paper and film is defined as

$$DR_{dB} = 10\log_{10}(D_{max} - D_{min}). \quad (12.3)$$

Printer transparency media are typically limited to a $D_{max}$ of ~2.0 and have a DR of ~17 dB. Paper media is limited to 22 dB or less. In both cases, higher values would be desirable.

Gamut defines the size of the CIE coordinate-defined color space. The gamut of a printer differs from that of a monitor (Fig. 12.9). The monitor color process is additive and the printer process is subtractive. The characteristics of printer ink differ from those of monitor phosphors. The gamut of a printer is not normally defined and is rather time consuming to determine. Several hundred color patches must be printed and measured with a colorimeter. A more practical solution is to use a test image containing a variety of saturated colors. Such an image is provided on the enclosed CD (fruit.tif).

Tonal transfer refers to the relationship between light intensity on the soft-copy display and on the hard-copy printer output. Because the display processes are totally different, there is no reason to assume the hard-copy output will look like the soft-copy display. Developing and implementing a printer compensation LUT is a key part of printer operation; it will be discussed in the following section. Tonal transfer also considers the issue of usable bit depth. Printer output is 8 bits per channel (256 levels) but, for a variety of reasons, 256 levels may not be discriminable. A 256-level step wedge printed after applying a printer compensation LUT can be used to measure usable bit depth. A rough approximation can also be made by printing the Briggs C-1 and C-3 targets, or any set of targets where a CL difference of 1–3 is portrayed.

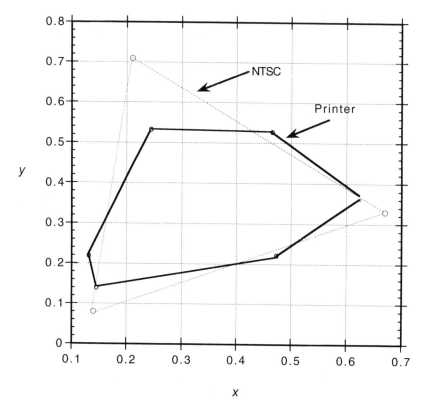

**Figure 12.9** Comparison of CRT and inkjet printer gamuts.

The task of matching hard-copy color to soft-copy color has been the subject of numerous studies. Details of the process are beyond the scope of this book, but more information can be found in Refs. [9] and [10]. Most color printers provide at least rudimentary color-matching software that defines a nominal monitor gamut and proceeds accordingly.

Large-scale uniformity refers to density variations across the format of the printed image. Uniformity was often a problem with earlier-generation thermal and photographic printers, but it is less of a significant issue with current-generation printers in good repair. Inkjet printers may occasionally show banding due to clogged inkjet nozzles. Laser printers may show banding due to variations in toner deposit.

Density uniformity can be evaluated by printing a full-format image at one or more CLs (e.g., 128, 50, 206). The resultant prints (or transparencies) can be measured using a densitometer or simply inspected. In the absence of a densitometer, a photometer can be used to measure transmittance of transparencies. The luminance of a flat-field light source (e.g., light box) is measured, and then the transparency is placed over the light source and intensity measured as a function of position. The photometer should be held stationary and the transparency moved. Transmittance is the ratio of output luminance (through the transparency) to input (measurement without the transparency). Transmittance measurements can be converted to density using Eq. (12.2).

Noise in the context of printers refers to high-frequency noise evident to the human observer. Since only photographic printers are continuous-tone devices, all printers will show tonal variation at some level of resolution. In the case of a halftone device, values alternate between the maximum and minimum, the mean and standard deviation (signal and noise) depend on the original image intensity value and the image/printer transfer function. The SNR values for the printer images shown in Fig. 12.8 are 10 dB (laser), 19 dB (inkjet), and 25 dB (thermal). The image mean and standard deviation were computed on the image of each printer output scanned at 1200 dpi. Depending on the application, the laser print may be perfectly acceptable. Therefore, it is suggested that rather than measure noise, it may be more expedient to simply observe sample outputs and make a perceptual judgement as to the adequacy of noise performance.

A type of noise that may be more of a problem is color misregistration. Thermal and laser printers produce color in three or four sequential passes; inkjet printers use one pass but with two cartridges and multiple nozzles. Any misregistration will result in color fringing or blurring of the image. A visual inspection at the normal viewing distance is suggested to "measure" the misregistration. One could, in theory, establish a threshold based on measures of visual acuity; however, the printing process is complex in that an image pixel is represented by several printer dots, and tones and colors are produced by varying the density and color of these dots. Further, the ability to perceive misregistration is a function of luminance and color (or CIE coordinate difference) and is thus not a single value. The enclosed CD provides a target (Regis.tif) that can be used to assess color misregistration, but it is recommended that users employ images representative of their own applications.

Geometric fidelity refers to the ability of the printer to accurately capture position or location. Because printers have moving parts in the writing process, the possibility exists for a nonlinear movement. In practice, however, this has generally not been seen as a problem beyond the previously mentioned issue of color misregistration.

As was the case with digitizers, volume and time must also be considered in selecting a printer. Inkjet printers trade quality for printing time and cost. Printing an 8-in. × 10-in. color image can take up to two minutes. Ink cartridges must be replaced relatively frequently, and photographic paper is more expensive than conventional bond paper. Costs of roughly $1 (U.S.) or more per print are typical. However, inkjet printers can be purchased for $300 or less. Thermal printers are also slow and relatively expensive per print (in the region of $2). The purchase cost is on the order of $5,000 to $10,000. Alternatively, laser printing is faster (several pages per minute) and less costly per print. The purchase costs range from $500 for monochrome printers to $2500 for color printers. For working prints and transparencies, a laser printer may be the best choice. For presentation print quality, inkjet or thermal printers may be needed. It should be remembered, however, that a reduction in image size and quality might be the trade-off. An image can be enlarged to effectively communicate its message by using a lower-quality printer as opposed to using a smaller enlargement on a higher-quality printer.

Finally, the issue of print longevity needs to be considered. Inkjet printers use water-soluble ink, so prints can be damaged or destroyed in the presence of water. They may also fade over time. A protective covering (e.g., lamination) can be applied to the print to protect it from water damage but at additional time and expense. Thermal prints also degrade over time, possibly involving color changes, even though a protective coating is applied. Laser prints offer the greatest longevity, but again, potentially at the expense of quality.

Most users will need multiple printer solutions. Photographic print quality can best be obtained with an inkjet or thermal printer. The inkjet solution is cheaper (and easier to operate), but print longevity is an issue. For high volume and speed, laser printers are superior, particularly with monochrome images. A low-cost inkjet and a moderate-cost monochrome laser printer represent a reasonable solution for most users. If a high volume of viewgraphs is to be produced, a color laser printer with a density range of $\geq 2.0$ would be desirable. Once a printer technology has been chasen, resolution (dpi) and DR are probably the most important selection factors. It should again be emphasized that resolution values (dpi) can be compared only within a printer technology; comparisons cannot be made across technologies.

### 12.2.3 Printing procedures

Printer operation begins with setup and calibration.[11] Inkjet printers typically provide a set of procedures designed to check print head alignment. Some printers are provided with a compensation LUT as well as settings for different media types. As a first step, alignment checks should be made and other vendor instructions followed. Where multiple options are provided, a nominal or "normal" setting should usually be used.

Aside from nominal setup and calibration, the single most important aspect of printer operation is the creation of a printer LUT. In some cases, the output of a printer will be found to be satisfactory using factory settings. However, optimized performance can be achieved by applying perceptual linearization procedures similar to those used for displays. The approach to developing a printer LUT is similar to that of developing a monitor LUT. First, the native response of the printer to CL variations must be determined. Normally, this is accomplished with a densitometer. In the absence of a densitometer, a photometer can be used to measure transparencies, as described in the previous section. Note that the printer response will vary with different media and must be remeasured each time a different print material is used. Even different batches of the same media can sometimes provide different results.

Once the native printer response is defined, the goal is to optimize or linearize the response in the same manner as was done for monitors. This requires defining a desired relationship between CL and density or transmittance, and then developing and applying a LUT to modify the native output to the desired output. Two methods are proposed. The first is the NEMA/DICOM recommended calibration,[12] the second is the Bartleson approach.[13]

The NEMA/DICOM method begins with a characterization of the printer response. A step wedge should be printed and measured with a densitometer. The step wedge should have ~64 steps ranging from CLmax to CLmin so as to accurately define the printer response. The response is likely to be nonlinear, as shown in Fig. 12.10. If the output is a transparency to be viewed on a light box (e.g., radiograph), the luminance of the light box should be measured. A correction for ambient light can be made by measuring the luminance of a typical image with the light box turned off. The measured contribution of room lighting is that which is reflected from the transparency. Density values from the measured step wedge are converted to transmittance values and multiplied by the luminance output of the light box. The contribution of room lighting is then added to the luminance value. Thus,

$$L_T = (0.1^D * L_{LB}) + L_A, \qquad (12.4)$$

where $L_T$ is the luminance at the measured density $D$, $L_{LB}$ is the luminance of the light box with no transparency, and $L_A$ is the additional luminance from room lighting.

If the output is a transparency to be projected, the luminance of the screen without a projected transparency should be measured and used in the same manner

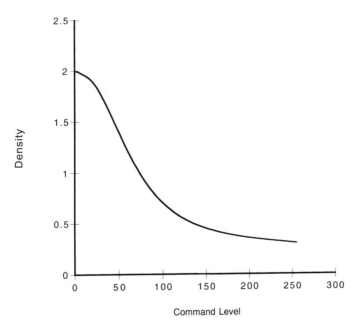

**Figure 12.10** Typical printer response.

as the luminance of the light box. It will be assumed that there is no room lighting other than the projector. Finally, if the output is a paper print, the reflected luminance of a minimum density paper should be measured under the lighting intended for viewing. This again represents the value for $L_{LB}$ in Eq. (12.4).

When the observed relationship between CL and displayed luminance is established, the desired relationship over the range of Lmin to Lmax should be determined using the procedures defined in Chapter 10. A LUT is then generated (again using the procedures in Chapter 10) to convert the observed printer output luminance function to the desired function.

The Bartleson method is based on the goal of maintaining a uniform perceptual brightness relationship across different methods of display. Bartleson determined that the shape of the brightness function changed with the viewing illumination conditions. Three levels were proposed—dark for movies and slides, dim for TV (and possibly displays), and bright for paper prints. Relative brightness was defined in terms of the CIE coordinate notation, where $Y$ represents luminance. The notation $Y/Y_o$ is commonly used to express "lightness," where $Y_O$ is maximum luminance. The ratio of two luminance values in the current context is equivalent to transmittance. Relative lightness (or brightness) is thus defined by transmittance. The aim functions (the mathematical functions or curves that one is "aiming" to achieve) for the three lighting conditions were defined by Bartleson as

$$L_{Bright} = 11.5(100T + 1)^{0.5} - 16, \qquad (12.5)$$

$$L_{Dim} = 17.7(100T + 0.6)^{0.41} - 16, \text{ and} \qquad (12.6)$$

$$L_{Dark} = 25.4(100T + 0.1)^{0.33} - 16. \qquad (12.7)$$

The aim curves are plotted in Fig. 12.11. The printer output density, $D$, is converted to transmittance by the relationship

$$T = 0.1^D. \qquad (12.8)$$

For transparencies, the dark equation [Eq. (12.7)] should be used to define relative luminance values. For paper prints, the bright equation [Eq. (12.5)] should be used for normal office lighting, and the dim equation [Eq. (12.6)] for subdued lighting. The values predicted by the equations represent the desired normalized transmittance/luminance relationship.

# DIGITIZERS, PRINTERS, AND PROJECTORS

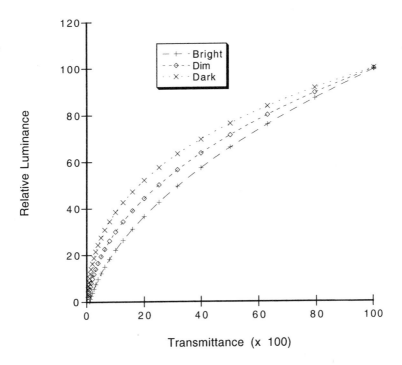

**Figure 12.11** Bartleson aim curves derived from Eqs. (12.5) to (12.7).

The process of defining the necessary Bartleson LUT is as follows:

1. Define the printer CL/density relationship.

2. Define the displayed print CL/transmittance function by applying Eq. (12.4) to the measured density values.

3. Invert the appropriate equation from Eqs. (12.5) through (12.7) to predict the desired value of $100T$ for a given desired normalized luminance value, $L$. The inverse of Eq. (12.7) is

$$100T = 10\frac{\log\left[\frac{L+16}{25.4}\right]}{0.33} - 0.1. \tag{12.9}$$

For the observed relationship, $100T$ equals the normalized value of $L$. Thus,

$$100T = L. \tag{12.10}$$

4. Invert the CL/transmittance function in step 2 above to define the relationship between CL and transmittance [CL = f(T)].

5. Use the relationship in step 4 with the equations from step 3 to predict CL for the full range (0–100) of normalized $L$ values for both the observed and the desired relationships.

6. Regress the desired CL values of the observed relationship to define a LUT for the printer. The LUT is to be applied to the image values before printing.

Ideally, this method should be applied only when the monitor has been properly calibrated and a linearization LUT applied. Since this method provides the same relative brightness as displayed on the monitor, any nonlinearities will be carried over to the print or transparency.

## 12.3 Projection Displays

Digital projection displays use optics to project a CRT or LCD image(s) onto a screen.[14] Both front and rear projection displays are used. The two most critical measures of performance are the resolution of the device (expressed in terms of addressability; e.g., 800 × 600) and light output. Light output is typically measured in lumens. A lumen is a measure of light falling on a defined area at a defined distance from the source. One lumen at a distance of 1 m illuminating a 1 m$^2$ area is equal to 1 lux. A projector with a large lumen value implies a large DR (which is desirable).

Many vendors advertise light output in terms of ANSI lumens. Light output is not constant across the projector field, so the American National Standards Institute (ANSI) developed a measurement procedure that accounts for this variation. This procedure requires that the maximum output of the projector be adjusted such that a 5% and a 95% patch are visually distinguishable from a 0%, 10%, 90%, and 100% patch. The process is repeated at nine locations over the screen and the results averaged. The method is subject to substantial uncertainty but is more conservative than measurement of an absolute maximum. The uncertainty makes comparison of data from different vendors difficult, so small differences should be ignored.

The resolution number for a projector simply indicates the size (in pixels) of the projected image. In general, all of the factors affecting the quality of a monitor also affect the quality of a projection device, but these factors are seldom measured. In addition, the quality and focus of the projection lens affect image quality. Because projection devices use cooling fans, noise can be an issue. Also, some CRT devices use a separate CRT for each of the three colors; optical alignment of the three beams is thus critical.

The guidance previously provided for printer LUTs also applies to projection devices. The CRT or LCD has a nonlinear response function that must be perceptually linearized for optimum quality. The I/O function must be defined and then a LUT generated using either the NEMA/DICOM method or the Bartleson approach.

## 12.4 Summary

Although digital image capture devices are becoming more prevalent, it may still be necessary to digitize images from a hard-copy source. A variety of digitizers are available to perform this task, differing in type, performance, and cost. Key performance parameters are resolution (the size of the digitizing spot) and radiometric fidelity. Other selection factors include the size and the volume of the material to be digitized.

The two key elements of the digitizing process are defining the digitizing resolution and making adjustments to the tonal/color transfer process. The goal of the tonal/color transfer process is to preserve all of the information in the original. The required resolution is based on how the digitized output will be viewed.

It may also be necessary to display a digital image in hard copy. Four major classes of printers are available for this task, differing in technology, cost, quality, and speed. Printer resolutions are generally adequate for unaided viewing (no optical magnification), so resolution is not a major selection issue. A more important distinction is the available density or DR—generally less than that of a soft-copy display. Nonlinearities in the printing process require the use of a printer compensation LUT. The same procedures used for perceptual linearization of displays can be applied to printers.

Finally, digital projection devices employ optics to project a CRT or LCD image on a front- or rear-projection screen. Light output is an important parameter. Resolution simply defines the number of pixels that can be displayed at one time. The perceptual linearization procedures used for monitors and printers are also applicable to projection displays.

## References

[1] J. Leachtenauer, K. Daniel, and T. Vogel, *Selecting and Using Image Digitizers*, National Exploitation Laboratory, Washington, DC (1995).
[2] Minolta Corporation, *3-D Non-Contact Scanning for the Medical Field*, Technical Application Brief, www.minoltausa.com/vivid/applications/pdfs/appsdental.pdf (2002).
[3] J. Leachtenauer, K. Daniel, and T. Vogel, "Preliminary comparison of digitizer quality metrics," in *Proceedings, Second International Airborne Remote Sensing Conference and Exhibition*, San Francisco, CA, Vol. III, pp. 83–91 (1996).
[4] J. C. Leachtenauer, K. Daniel, and T. P. Vogel, "Digitizing corona imagery: quality vs. cost," in R. A. McDonald (ed.), *Corona Between the Sun and the Earth: The First NRO Reconnaissance Eye in Space*, ASPRS, Bethesda, MD, pp. 189–203 (1997).
[5] Corel Corporation, *Commercial Printing Guide,"* Ontario, Canada (1998).

[6] J. C. Leachtenauer and R. G. Driggers, *Surveillance and Reconnaissance Imaging Systems: Modeling and Performance Prediction*, Artech House, Boston, MA (2001).

[7] RIT Research Corp, *On Demand Printers, A Product Survey*, Rochester, NY (1996).

[8] National Exploitation Laboratory, *Selecting an Image Printer: Factors to Consider*, NPIC, Washington, DC (1994).

[9] H. R. Kang, *Color Technology for Electronic Imaging Devices*, SPIE Press, Bellingham, WA (1996).

[10] E. Baumann and R. Hofmann, "Color aspects in photo-quality ink-jet printing," *IS&T Reporter*, Vol. 17(3), pp. 1–4 (2002).

[11] Central Imagery Office, *USIS Standards and Guidelines*, CIO-2008, Vienna, VA (1995).

[12] National Electrical Manufacturers Association, *Digital Imaging and Communications in Medicine (DICOM) Part 14: Grayscale Standard Display Function*, Rosslyn, VA (2001).

[13] C. J. Bartleson, "Optimum image tone reproduction," *Journal of the SMPTE*, Vol. 84, pp. 613–618 (1975).

[14] P. A. Keller, *Electronic Display Measurement*, John Wiley & Sons, Inc., New York (1997).

# Appendix
# Test Targets

This appendix describes the test targets provided on the attached CD. The monochrome targets are all 8-bit images (1600 × 1200 pixels) in size. Color targets are 8-bits per channel (RGB) and also 1600 × 1200 pixels. Table A-1 lists each target, its application, and the section describing its use. It is recommended that the images be viewed with an image viewer that does not modify the images. A program called Image J provides this capability. Image J is freeware available from http://rsb.info.nih.gov/ij/ that runs on Windows, Linux, Unix, OS-2, and Macintosh.

Some of the targets require uniform distribution across the face of the monitor. The target used to measure addressability (Address.tif) is an example. For monitors with other than 1600 × 1200 addressability, adjustment will be required. Similarly, some targets (e.g., Uni75.tif) require a CL resulting in a luminance output at a defined percentage of Lmax. Again, adjustment will be required. Finally, the Cm targets assume a 1600 × 1200 monitor at 100 ppi.

**Table A-1** Test targets.

| Category | File | Description | Application | Reference |
|---|---|---|---|---|
| Briggs target | BrgC1.tif | Monochrome Briggs with 1 count contrast | Display quality (Cm) | Sec. 5.4 |
| | BrgC3.tif | Monochrome Briggs with 3 count contrast | Display quality (Cm) | Sec. 5.4 |
| | BrgC7.tif | Monochrome Briggs with 7 count contrast | Display quality (Cm) | Sec. 5.4 |
| | BrgC15.tif | Monochrome Briggs with 15 count contrast | Display quality (Cm) | Sec. 5.4 |
| | BrgYG.tif | Color Briggs—Yellow/green | Color display quality | Sec. 5.4 |
| | BrgBG.tif | Color Briggs—Blue/green | Color display quality | Sec. 5.4 |
| | BrgYR.tif | Color Briggs—Yellow/red | Color display quality | Sec. 5.4 |
| Luminance measurement | Setup.tif | CL 1 box in CL 0 background | Dark cutoff | Sec. 7.4 |
| | IOFunc0.tif | Box at CL=0 in background | Define I/O function | Table 7.4 |

**Table A-1** (cont.) Test targets.

| Category | File | Description | Application | Reference |
|---|---|---|---|---|
| Luminance measurement (cont.) | IOFunc1.tif | Box at CL=1 in background | Define I/O function | Table 7.4 |
| | IOFunc2.tif | Box at CL=2 in background | Define I/O function | Table 7.4 |
| | IOFunc3.tif | Box at CL=3 in background | Define I/O function | Table 7.4 |
| | IOFunc4.tif | Box at CL=4 in background | Define I/O function | Table 7.4 |
| | IOFunc6.tif | Box at CL=6 in background | Define I/O function | Table 7.4 |
| | IOFunc8.tif | Box at CL=8 in background | Define I/O function | Table 7.4 |
| | IOFunc11.tif | Box at CL=11 in background | Define I/O function | Table 7.4 |
| | IOFunc16.tif | Box at CL=16 in background | Define I/O function | Table 7.4 |
| | IOFunc23.tif | Box at CL=23 in background | Define I/O function | Table 7.4 |
| | IOFunc32.tif | Box at CL=32 in background | Define I/O function | Table 7.4 |
| | IOFunc45.tif | Box at CL=45 in background | Define I/O function | Table 7.4 |
| | IOFunc64.tif | Box at CL=64 in background | Define I/O function | Table 7.4 |
| | IOFunc91.tif | Box at CL=91 in background | Define I/O function | Table 7.4 |
| | IOFunc128.tif | Box at CL=128 in background | Define I/O function | Table 7.4 |
| | IOFunc181.tif | Box at CL=181 in background | Define I/O function | Table 7.4 |
| | IOFunc255.tif | Box at CL=255 in background | Define I/O function | Table 7.4 |
| | CLmax.tif | Full screen white | Lmax, viewing angle | Sec. 7.4 |
| | CLmin.tif | Full screen black | Lmin, DR, viewing angle | Sec. 7.4 |
| | Halat.tif | Clmin box in Clmax background | Measure halation | Sec. 7.4 |
| | Uni75.tif | Full screen at 75% Lmax | Measure uniformity | Sec. 7.5 |

**Table A-1** (cont.) Test targets.

| Category | File | Description | Application | Reference |
|---|---|---|---|---|
| Luminance measurement (cont.) | Uni50.tif | Full screen at 50% Lmax | Measure uniformity | Sec. 7.5 |
| | Uni25.tif | Full screen at 25% Lmax | Measure uniformity | Sec. 7.6 |
| | Unif.tif | Step wedge pattern | Assess luminance uniformity | Sec. 7.5 |
| | Angle.tif | CLmax & CLmin square in background | Assess viewing angle | Sec. 7.5 |
| Color measurement | Colors.tif | Defined colors | Assess color | Table 2.1 |
| | Fruit.tif | Image of basket of fruit | Assess color | Sec. 7.5 |
| | RGB.tif | Squares at Rmax, Bmax, and Gmax | Assess color | Sec. 7.5 |
| | Cunif.tif | Gray squares at different intensities | Assess color uniformity | Sec. 7.5 |
| | Regis.tif | CMY lines and squares | Assess printer registration | Sec. 12.2 |
| Size/resolution measurement | Aspect.tif | 400 pixel square | Measure aspect ratio | Sec. 8.4 |
| | Address.tif | Grill plus outlines and diagonals | Measure addressability | Sec. 8.4 |
| | Density.tif | Equally spaced line pattern | Measure pixel density | Sec. 8.4 |
| | CmAVert.tif | Grill patterns | Measure vertical Cm in Zone A | Sec. 8.4 |
| | CmAHor.tif | Grill patterns | Measure horizontal Cm in Zone A | Sec. 8.4 |
| | CmBVert.tif | Grill patterns | Measure vertical Cm in Zone B | Sec. 8.4 |
| | CmBHor.tif | Grill patterns | Measure horizontal Cm in Zone B | Sec. 8.4 |
| | Cmgrill.tif | Grill pattern | Assess Cm | Sec. 8.5 |
| | Cm.tif | Grill patterns on various backgrounds | Assess Cm | Sec. 8.5 |
| Noise/artifacts/ distortion | JitterV.tif | Horizontal grill | Measure jitter/swim/drift | Sec. 9.4 |
| | JitterH.tif | Vertical grill | Measure jitter/swim/drift | Sec. 9.4 |

**Table A-1** (cont.) Test targets.

| Category | File | Description | Application | Reference |
|---|---|---|---|---|
| Noise/artifacts/ distortion (cont.) | Microjit.tif | Horizontal/vertical hash marks | Measure micro jitter | Sec. 9.4 |
| | Step.tif | 90% box in 10% background | Measure step response | Sec. 9.4 |
| | Mura.tif | 50% field | Determine mura | Sec. 9.4 |
| | Ghost.tif | Grill targets | Determine ghosting | Sec. 9.4 |
| | LinearV.tif | Vertical full-screen grill | Measure linearity | Sec. 9.4 |
| | LinearH.tif | Horizontal full-screen grill | Measure linearity | Sec. 9.4 |
| | Distort.tif | Full-screen grid | Assess geometric distortion | Sec. 3.7 |

# Index

β, 99
$\Delta E_{IL}$, 133
4-AFC metric, 94

## A

acceptance angle, 16
accommodation, 107
achromatic, 29
acoustical noise, 209
ACR Standard for Teleradiology, 143
active-matrix liquid crystal display,
　see AMLCD
adaptation, 105
additive color, 17, 40
addressability
　and screen size, 54
　as part of measurement definition, 162
　consideration in equipment selection, 201
　control, 44
　defined, 33
　minimum, 166
　procedure for measuring, 171
addressable pixels, 30
aerial imagery, 143
age
　of monitor, 74
　of observer, 121
aim curves, 256
aim functions, 256
aliasing, 75
alternating current, 43
alternatives to measurement, 194
ambient light, 16, 204–205, 208
AMLCD, 29, 41
amplified beam current, 34
amplified voltage, 34
amplitude, 31
analog
　device, 29
　signal, 31
　TV, 31
angular disparity, 110
annotation capability, 199
ANSI lumens, 258
antireflection coating, 38
application-specific integrated circuits, 201
array scanners, 237
artifacts, 53, 74, 184, 186, 192

astigmatism, 121
average signal level, 71

## B

banding, 252
bandwidth, 229
bandwidth compression, 219, 229
barrel, 203
Barten contrast sensitivity model, 126
Bartleson, 255
beta (β), 99
bicubic interpolation, 228
bilinear interpolation, 163, 228
binary decision, 84
binocular vision, 110
bit depth, 33, 145, 153, 241
blackbody, 49
Blackwell, R., 124
blemishes, 75
blurring, 253
bone cortex, 231
Briggs
　color ratings, 140
　scores, 91, 95, 212
　target, 88–89, 93
brightness, 15, 17, 44
brightness nonuniformities, 184

## C

calibration
　of printer, 254
　source, 17
camera exposure meter, 16
candelas, 13
capture device, 1, 6
cathode, 36
cathode-ray tube, see CRT
CCD
　array, 17
　response to light energy, 238
CD accompanying book, contents of, 8
center area of monitor, 62
charge-coupled device, see CCD
checkerboards, 89
checkers, 90
chest images, 142, 231
chroma, 23
CIE (Commission Internationale de l'Eclairage), 17

265

CIE L*a*b*, 133
CL, see command level
CL/luminance function, 211
clock frequency, 201
clutter, 130
Cm, see contrast modulation
CMYK, 25
cognitive measures, 8
cognitive system, 81
coherent illumination, 130
color
    bit depth, 44
    blindness, 123
    Briggs, 140
    Briggs target, 93
    casts, 38, 147
    constancy, 110
    controls, 44
    CRT, 40
    differences, 117
    fringing, 253
    gain, 44, 49
    laser printers, 249
    matching, 252
    measurement, 11, 17
    misregistration, 253
    monitors, 44, 54, 70, 137
    purity, 40
    temperature, 44, 48, 139, 154, 147
    tracking, 70
    transforms, 223
    uniformity, 155
    vision deficiencies, 123
    vs. monochrome monitor selection, 200
colorfulness, 23
colorimeter, 26
colors displayed, number of, 49
command level (CL)
    defined, 22
    and luminance, 211–212
computed radiography, 231
computed tomography (CT), 1, 145
computer memory (CPU), 34
cones, 107
confidence ratings, 84
contouring, 145
contrast, 44, 53
contrast modulation (Cm), 61, 162, 166
    Zone A, 172
    Zone B, 172
contrast ratio, 66
contrast sensitivity, 112
controller, 201
convergence, 56
convolution kernel, 224
coordinate system, 17
cornea, 107

corrective lenses, 121
correlation between CL and luminance, 212
correlation coefficient, 163
criteria scaling, 85
Crites, 212
CRT
    as display type, 29
    color, 39–40
    flat panel, 39
    operation
        efficiency of, 38
        role of electrons in, 34
CT, see computed tomography
cycle, 57

# D

$d'$, 99
decibels (dB), 138
degaussing control, 209
delta pulse code modulation, 230
density uniformity, 252
desired values, 138
detective quantum efficiency (DQE), 74
diagonal, 54
diagonal pitch, 173
dichromats, 123
differential receiver operating characteristic (DROC), 101
diffuse reflection, 68
digital brightness and contrast, 201
digital frequency, 31
digital signal, 31
digital-to-analog converter, 138
digitizer, 237
    device selection, 239
    image quality, 239
    operation, 237
    resolution, 240
digitizing, 237
    procedures, 243
    resolution, 244–245
direct optical projection, 248
discrete cosine transform, 230
dispenser cathode, 36
display
    controller, 5, 31
    frequency control, 44
    quality metrics, 8
    system, 5
distortions, 53, 74
Doppler ultrasound, 223
dot density, 244
double stimulus continuous quality scale, 83
double stimulus impairment scale, 83
DQE, see detective quantum efficiency

# INDEX

DR
    adjustment (DRA), 219–220
    defined, 66
    to measure luminance, 138, 149
    to measure radiometric fidelity, 240, 251
DRA, *see* DR adjustment
drift, 74, 181, 183, 189
DROC, *see* differential receiver operating characteristic
drum microdensitometers, 237
dye
    diffusion, 249
    sublimation, 249
dynamic black-level stability, 69
dynamic focusing, 36
dynamic range, *see* DR

## E

edge
    response, 60
    sharpness, 59
efficiency of CRT operation, 38
egg crate diffusers, 207
electron gun, 36
electronic projection devices, 2
electrons, role in CRT operation, 34
electrostatic lens system, 36
electrostatic printers, 249
emitted energy, 11
emulsion, 243
environmental controls, 199
ergonomic considerations, 209
etching, 38
event distributions, 99
exposure
    meter, 155
    value, 155
extinction
    level, 184
    ratio, 182, 192

## F

face plate, 38
far-sightedness, 121
fast Fourier transform, 73
fast intelligent tracking, 40
FEDs, *see* field-emissive displays
field of view, 16
field-emissive displays (FEDs), 29
film
    annotation and loading, 243
    digitizers, 1
    resolution, 240
fixation, 102, 109
flatbed microdensitometers, 237
flatbed linear scanners, 238
flat-panel displays, 29
flicker, 74, 183
flicker mechanism, 110
footcandle, 13
foot-lamberts defined, 13
fovea, 107
foveal vision, 109
frame grabbers, 237
free-response operating characteristic (FROC), 101
frequency, 31
frequency filtering, 227
FROC, *see* free-response operating characteristic

## G

gamma, 34, 66, 147
gamut, 22, 70, 148, 251
geometric correction, 242
geometric distortion, 75
    controls, 44
geometric fidelity, 239, 242, 249
geometric transforms, 219, 227
ghosting, 75, 184
graphics acceleration, 34
graphics cards, 31
gray-level reversal, 71
gray-scale images, 223
grid, 36

## H

halation, 67, 139, 145, 153
half-power level, 56
halftone printing, 244, 250
hard-copy imagery, 1
haze component, 68
HDTV, *see* high-definition TV
high-definition TV (HDTV), 30
high-frequency boost, 224
horizontal pitch, 173
HSB space, *see* hue/saturation/brightness space
hue
    defined, 17
    differences, 24
    with color imagery, 223
hue/saturation/brightness (HSB) space, 24
human observer, 3
human visual system (HVS), 6, 8, 105
    individual differences in, 119
hyperspectral, 231

## I

IDEX
    function, 214
    system, 125

illuminance, 13
illuminance photometer, 15
image
 chain, 1
 compression, 230
 interpretability, 85
 motion, 115
 roam, 182
 scaling, 85
Image Display and Exploitation system, *see* IDEX
Imagery Interpretability Rating Scale, 81
impedance, 201, 203
individual differences in HVS, 119
information technology (IT) personnel, 7
information theory, 187
initial setup, 149
initialization and prescan, 243
inkjet printers, 249
input/output function, *see* I/O function
instantaneous signal level, 71
Integrated Exploitation Capability, 137
interface requirements, 201
interlaced scan, 36
internal noise, 126
interpolated resolution, 240
I/O function, 42, 67, 147, 150
ionized gas, 43
iris, 105

## J

"J" curve, 111
jitter, 74, 181–183, 189–190
JPEG, 230
JPEG 2000, 230

## K

kernel, 226

## L

$L^*a^*b^*$ space, 22
$L^*u^*v^*$, 20, 133
LaGrange interpolator, 228
landscape monitor mode, 30
large-scale uniformity, 240–241, 251
laser digitizers, 238
lateral disparity, 111, 117
lateral inhibition, 126
lens of human eye, 107
lens flare, 148
lesion margins, 231
light
 box, 141, 256
 energy, 37
 level, 2
 measurement of, 11
 output, 258
 sensitivity, 109
lightness, 223
linear TTA, 221
linearity, 46, 183, 188, 194, 240
liquid crystals, 41
Lmax, 139, 150
Lmin, 139
logarithmic amplifier, 238
look-up tables, 67
low-pass filter, 226
lumen, 11
luminance, 53, 77
 controls, 44
 JNDs, 139, 146
 loading, 69
 nonuniformity, 69, 144
 parameters, 137
 requirements
  color, 138
  monochrome, 139
  stereo, 139
 stability, 69, 147
 step response, 182, 190
 uniformity, 69, 150
luminescent phosphor, 34
luminous efficiency function, 21
luminous power, 11
lux, 13

## M

MacAdam ellipses, 20
macro jitter, 182, 190
magnetic
 environments, 200
 fields, 148, 200, 208
 resonance imaging, *see* MRI
magnification, 227
magnitude, 31
mammograms, 143, 212
mammographic reading rooms, 206
marker images, 85
matching, 252
matrix (CCD array) photometer, 173
matrix displays, 43, 73
maximum luminance, 139
measurement
 alternatives, 194
 domains, 53
 pucks, 16
mesopic vision, 115
micro jitter, 182, 190
microdensitometer, 237–238
microphotometer, 17

# INDEX

milli-lambert, 15
minification, 227
minimum addressability, 166
minimum performance levels, 138
modulation threshold, 113, 213
modulation transfer function, *see* MTF
moiré, 75, 182–183, 191
monitor
  aging of, 74, 144
  and video controller selection, 199
  calibration, 209
  center area of, 62
  color, 44, 54, 70, 137
  connection, 203
  environment, 204
  landscape mode, 30
  monochrome, 29, 137
  peripheral area of, 62
  portrait mode, 30
  setup, 203
  size/resolution requirements, 161–162
  surround, 208
  type selection, 199–200
monochromatic vision, 123
monochrome monitors, 29, 137
monochrome (luminance) models, 124
monocular vision, 110
motion imagery, 232
MPEG-1, 232
MPEG-2, 232
MPEG-4, 232
MRI, 145
  scan, 3
MTF, 63
  compensation, 224
  of the eye, 113
MTFC, 224
multiple-purpose monitor, 199
multispectral
  imagery, 212
  NIIRS, 140
  scanners, 1
multisync, 199
Munsell system, 17
mura, 75, 182, 184, 192
myopes, 120

## N

National Imagery and Mapping Agency,
  *see* NIMA
National Imagery Interpretability Rating Scale, *see*
  NIIRS
National Information Display Laboratory, *see* NIDL
nearest-neighbor resampling, 228
near-lossless archival storage, 243
near-sightedness, 119

NEMA/DICOM, 212
  calibration, 206
  perceptual linearization function, 128
neon glow lamps, 43
nerve cells, 109
neural impulses, 109
neural network processing, 1
NIDL, 17, 137
NIIRS, 2, 85
  ratings, 95
  variability, 87
NIMA, 137
NIST, 17
nit, 15
noise
  adjustment control, 47
  generally, 53, 71, 240–241, 251, 253
  types of
    acoustical, 209
    frequency-dependent, 186
    pink, 186
    spatial, 186
    temporal, 186
    time-independent, 186
    white, 186
noise-equivalent quanta, 74
noise power, 73
noise power spectrum, 74
noncoherent illumination, 130
noncommandable pixels, 42
nonlinear TTA, 221
nonuniform energy distribution, 56
nonuniformities, 184

## O

objective perceptual quality measures, 87
objective performance measures, 96
objective quality measures, 82
optical resolution, 240
organic light-emitting diodes, 29
overshoot, 69
oxide cathode, 36
oxide coating, 36

## P

paper print, 256
PDP, 29, 43
perceptible acuity, 111
perception of depth, 110
perceptual linearization, 4, 67, 201, 210
perceptual measures, 8, 81
performance measurement alternatives, 194
peripheral area of monitor, 62
peripheral vision, 109
periphery of retina, 107

phosphor
    in monitor, 34
    persistence of, 37
    SNR differences, 188
photometer, 15, 149
photometry, 11
photomultiplier tube, 238
photopic vision, 115
physical domain, 53
physical measures, 53
physically realizable colors, 18
picture archiving and communication, 207
pincushion, 203
pixel
    aspect ratio, 162, 170
    decimation, 163
    defects, 182–183, 193
    density, 55, 162–163, 171
    fill factor, 57
    intensity transforms, 219
    pitch, 55
    processing, 219
        operations, 233
        transforms, 1
    replication, 163, 228
    size, 55
    subtense, 57
plasma display panels, *see* PDP
point source, 11
polarization, 42, 111
polarizers, 41
polynomial, 211
portrait monitor mode, 30
position controls, 47
power, as unit of measure, 11
power saver, 215
presbyopes, 121
primary colors, 17
primary diagnosis, 199
printer
    calibration, 254
    look-up table (LUT), 254
    operation, 249
    quality, 249
    selection, 249
    setup, 254
printing procedures, 254
prints, 1
prismatic diffuser, 207
probability of detection, 100
probability of false alarms, 101
processor, 5
progressive scan, 36
projection displays, 258
protective covering of prints, 254
pseudocolor, 49, 133, 223
pulmonary nodules, 142

pupil diameter, 105
purchasing advice, 8

## Q

quality, 6, 29, 53
quantization, 212
quantizing, 32

## R

radar imagery, 206, 213
radiant energy, 11, 34
radiograph, 1, 144
radiography, 145
radiologists, 142
radiometric fidelity, 239–240, 249, 251
radiometry, 11
random access memory (RAM), 34
resolution-addressability ratio (RAR), 59, 175
raster
    modulation, 62
    pattern, 48
    stability, 77
recalibration, 215
receiver operating characteristic, 100
reconnaissance and surveillance community, 2
reflectance, 68, 147
reflected light, 15
refresh rate, 33, 48, 74, 181–183, 201
relative edge response, 59
resolution, 53, 258
    and cost trade-off, 243
    conversion, 240
    reduced, 233
resolution-addressability ratio, 59
resolvable pixels, 61
retina, 105
    periphery of, 107
retinal illuminance, 105
RGB model, 24
ringing, 75, 182–183
roam, 227
rods, 107
Rogers and Carel, 168, 212
room lighting, 2
rotation, 227

## S

saccades, 102, 109
saccadic distances, 110
saturation, 17, 23, 223
scan rate, 189
scanner, selection of, 237
scanning, 243

photometer, 173
process, 102
spectroradiometers, 26
scotopic vision, 115
screen
   aspect ratio, 162, 170
   buffer, 34
   saver, 215
   size, 54, 162, 170
search
   pattern, 102
   process, 102
   task, 102
sensitivity function, 11
separable acuity, 111
sequential color separation, 40
setup of printer, 254
shadow mask, 40
shielding, 209
signal, 99
   amplitude, 34
   plus noise, 99
signal-to-noise ratio, *see* SNR
silver halide printers, 249
size/resolution requirements
   of color monitor, 161
   of monochrome monitor, 162
slide scanners, 238
SMPTE target, 202
Snellen letter, 87
SNR, 71, 182, 193
solid angle, 13
spatial, 53, 77
   fidelity, 239, 249
   filtering, 224
   filters, 219
   integration process, 126
speckle, 130
spectral, 77
   color, 53
   sensitivity, 11
spectroradiometer, 26
specular reflections, 68
spot growth, 140
spot size enlargement factor, 247
standard sensitivity function, 12
steradian, 13
stereo acuity, 111, 117
stereoscopic, 110
Stiles-Crawford effect, 107
storage
   and processing device, 5
   and verification, 243
straightness, 75, 183, 188, 193
subjective
   performance, 82
   quality ratings, 83

quality scale, 85
ratings, 81
subtractive color concept, 25
supertwisted nematic (STN), 42
swim, 74, 181, 183, 189
synthetic aperture radar, 130, 224

## T

tasks, 6
temperature, 11
temporal, 53, 77
theory of signal detection (TSD), 2, 99
thermal printers, 249
TIFF, 248
timing conflicts, 203
tonal color transfer, 251
tonal transfer adjustment (TTA), 220
   linear, 221
   nonlinear, 221
tonal transfer correction (TTC), 220
transmittance, 68
transparency, 255
tri-bar, 87
trichromatic theory of vision, 110
tristimulus values, 18
troland, 105
TSD, *see* theory of signal detection
TTA, *see* tonal transfer adjustment
TTC, *see* tonal transfer correction
twisted nematic (TN), 41
two-dimensional FFT, 226

## U

undershoot, 69
Uniform Chromaticity Spacing (UCS), 20
uniform perceptual brightness, 256
unsharp masking, 226
utility of displayed image, 53, 81

## V

validation, 87
value of displayed image, 53
vernier acuity, 111
vertical grill, 40
vertically slotted mask, 40
VGA refresh rate, 33
video
   amplifier, 203
   card, 5
   controllers, 31
viewgraphs, 1
viewing, 110
   angle, 71

angle threshold, 139, 145, 152
    distance, 57
visible spectrum, 11
vision
    mesopic, 115
    photopic, 115
    scotopic, 115
    trichromatic theory of, 110
    *see also* HVS, individual eye parts
visual acuity, 111
visual angle, 57
voltage, 34

## W

warm-up time, 74, 148, 181, 188
warp, 227
water-soluble ink, 254
watt, 11

wavefront aberrations, 121
wavelength, 11, 17
    band, 11
    sensitivity, 11
    spectrum, 11
wavelet compression, 230
waviness, 75, 193
white point, 22

## X

x ray, 1
xenon, 43
XGA refresh rate, 33

## Z

Zone A, 162
Zone B, 162

Jon Leachtenauer received his A.B. and M.S. degrees in Geology from Syracuse University. After serving as a photointerpreter in the U.S. Army, he began a 40-year career in human factors research in image quality and image interpretation performance measurement. He has worked for Aero Service Corporation, Photics Research Corporation, the Boeing Company, and ERIM. While at ERIM, he was the Senior Scientist for the Imagery Analysis Division of the National Exploitation Laboratory. He formed a consulting company in 1999 (J/M Leachtenauer Associates) and is currently a consultant to the National Imagery and Mapping Agency and the U.S. Army Night Vision and Electronic Sensors Directorate. He also conducts research in image and display quality and medical automation. With Ronald Driggers, he is the author of *Surveillance and Reconnaissance Imaging Systems: Performance Prediction and Modeling*. He is a contributor to the *Encyclopedia of Optical Engineering* and other books, and is responsible for numerous technical reports and published papers. He is a member of SPIE, SID, OSA, Sigma Xi, and IS&T.